Douglas Quadling and Julian Gilbey

Cambridge International AS and A Level Mathematics:
Mechanics 2

Coursebook

Revised Edition

CAMBRIDGE
UNIVERSITY PRESS

CAMBRIDGE
UNIVERSITY PRESS

University Printing House, Cambridge CB2 8BS, United Kingdom

One Liberty Plaza, 20th Floor, New York, NY 10006, USA

477 Williamstown Road, Port Melbourne, VIC 3207, Australia

4843/24, 2nd Floor, Ansari Road, Daryaganj, Delhi – 110002, India

79 Anson Road, #06–04/06, Singapore 079906

Cambridge University Press is part of the University of Cambridge.

It furthers the University's mission by disseminating knowledge in the pursuit of education, learning and research at the highest international levels of excellence.

Information on this title: education.cambridge.org

First published 2002
Second edition 2016
20 19 18 17 16 15 14 13 12 11 10 9 8 7 6 5 4 3

Printed in Great Britain by CPI Group (UK) Ltd, Croydon CR0 4YY

A catalogue record for this publication is available from the British Library

ISBN 978-1-316-60033-7 Paperback

Image credits: Cover plainpicture/Westend61/Martin Rietze; Chapters 1, 3, 8 agsandrew/Shutterstock; Chapter 2 tmwwt/Shutterstock; Chapters 4, 6 TitimaOngkantong/Shutterstock; Chapter 5 bdavid32/Shutterstock; Chapter 7 xiaoke ma/Getty Images; Chapter 9 appler/Shutterstock; Chapter 10 GrandeDuc/Shutterstock; Photo in Chapter 6, exercise 6B, question 9 Zoonar GmbH/Alamy Stock Photo

Contents

Contents

Introduction

Cambridge International AS and A Level Mathematics has been written especially for the Cambridge International Examinations syllabus. There is one book corresponding to each syllabus unit, except that units P2 and P3 are contained in a single book. This book covers the second Mechanics unit, M2.

The syllabus content is arranged by chapters which are ordered so as to provide a viable teaching course. Occasionally a section or an example deals with a topic which goes beyond the requirements of the syllabus; these are marked with a vertical coloured bar.

Chapters 7 and 10, on geometrical methods and on strategies, do not introduce any essentially new principles, but offer alternative methods for dealing with certain types of problem to those given in earlier chapters. They have been included primarily for students aiming at higher grades. Knowledge of the material in these chapters will not be required in the examination, but may suggest neater methods of solution. The vector approach to projectile motion in Chapter 1, and the associated notation, is also not an examination requirement.

The treatment of force as a vector quantity introduces some simple vector notation, consistent with that in Chapter 13 of unit P1. Students may find this useful in writing out their solutions, although it is not a requirement of the examination.

Some paragraphs within the text appear in *this type style*. These paragraphs are usually outside the main stream of the mathematical argument, but may help to give insight, or suggest extra work or different approaches.

Numerical work is presented in a form intended to discourage premature approximation. In ongoing calculations inexact numbers appear in decimal form like 3.456..., signifying that the number is held in a calculator to more places than are given. Numbers are not rounded at this stage; the full display could be, for example, 3.456 123 or 3.456 789. Final answers are then stated with some indication that they are approximate, for example '1.23 correct to 3 significant figures'.

The approximate value of g is taken as $10\,\mathrm{m\,s^{-2}}$.

There are plenty of exercises, and each chapter ends with a Miscellaneous exercise which includes some questions of examination standard. There are two Revision exercises, and two Practice Exam-style papers.

Some exercises have a few questions towards the end which are longer or more demanding than those likely to be set in the examination. Teachers may wish to assign these questions to selected students to enhance the challenge and interest of the course. Questions marked with a vertical Coloured bar require knowledge of results outside the syllabus.

The author thanks Cambridge International Examinations and Cambridge University Press for their help in producing this book. However, the responsibility for the text, and for any errors, remains with the author.

Douglas Quadling, 2002

Introduction to the revised edition

This revised edition has been prepared to bring the textbook in line with the current version of the Cambridge International Examinations specification. As much as possible of the original edition has been left unchanged to assist teachers familiar with the original edition; this includes section numbers, question numbers and so on. A short discussion on the limitations of the projectile model has been added at the end of section 1.1.

The major change in this edition is the replacement of all of the older OCR examination questions in the Miscellaneous Exercises by more recent Cambridge International Examinations questions. This will be of benefit to students preparing for the current style of examination questions. In order to maintain the numbering of the other questions, the newer questions have been slotted in to the exercises. While this has inevitably meant some loss of order within the Miscellaneous Exercises, this was felt to be more than compensated for by the preservation of the original numbering. In a few of the exercises, an insufficient number of past paper questions were available to replace the existing questions; in these cases, the exercises have either been shortened or new questions written to replace the old ones. Further past papers can, of course, be requested from Cambridge International Examinations. All questions and answers taken from Cambridge International Examinations past papers have been clearly referenced. All other questions and answers have been written by the authors of this book.

The editor of this edition thanks Cambridge International Examinations and Cambridge University Press, in particular Cathryn Freear and Andrew Briggs, for their great help in preparing this revised edition.

Julian Gilbey

London, 2016

Chapter 1
The motion of projectiles

In this chapter the model of free motion under gravity is extended to objects projected at an angle. When you have completed it, you should

- understand displacement, velocity and acceleration as vector quantities
- be able to interpret the motion as a combination of the effects of the initial velocity and of gravity
- know that this implies the independence of horizontal and vertical motion
- be able to use equations of horizontal and vertical motion in calculations about the trajectory of a projectile
- know and be able to obtain general formulae for the greatest height, time of flight, range on horizontal ground and the equation of the trajectory
- be able to use your knowledge of trigonometry in solving problems.

Any object moving through the air will experience air resistance, and this is usually significant for objects moving at high speeds through large distances. The answers obtained in this chapter, which assume that air resistance is small and can be neglected, are therefore only approximate.

1.1 Velocity as a vector

When an object is thrown vertically upwards with initial velocity u, its displacement s after time t is given by the equation

$$s = ut - \tfrac{1}{2}gt^2,$$

where g is the acceleration due to gravity.

One way to interpret this equation is to look at the two terms on the right separately. The first term, ut, would be the displacement if the object moved with constant velocity u, that is if there were no gravity. To this is added a term $\tfrac{1}{2}(-g)t^2$, which would be the displacement of the object in time t if it were released from rest under gravity.

You can look at the equation

$$v = u - gt$$

in a similar way. Without gravity, the velocity would continue to have the constant value u indefinitely. To this is added a term $(-g)t$, which is the velocity that the object would acquire in time t if it were released from rest.

Now suppose that the object is thrown at an angle, so that it follows a curved path through the air. To describe this you can use the vector notation which you have already used (in M1 Chapter 10) for force. The symbol **u** written in bold stands for the velocity with which the object is thrown, that is a speed of magnitude u in a given direction. If there were no gravity, then in time t the object would have a displacement of magnitude ut in that direction. It is natural to denote this by **u**t, which is a vector displacement. To this is added a vertical displacement of magnitude $\tfrac{1}{2}gt^2$ vertically downwards. In vector notation this can be written as $\tfrac{1}{2}$**g**t^2, where the symbol **g** stands for an acceleration of magnitude g in a direction vertically downwards.

To make an equation for this, let **r** denote the displacement of the object from its initial position at time $t = 0$. Then, assuming that air resistance can be neglected,

$$\mathbf{r} = \mathbf{u}t + \tfrac{1}{2}\mathbf{g}t^2.$$

In this equation the symbol + stands for vector addition, which is carried out by the triangle rule, the same rule that you use to add forces.

EXAMPLE 1.1.1

A ball is thrown in the air with speed $12\,\mathrm{m\,s^{-1}}$ at an angle of $70°$ to the horizontal. Draw a diagram to show where it is 1.5 seconds later.

If there were no gravity, in 1.5 seconds the ball would have a displacement of magnitude 12×1.5, that is $18\,\mathrm{m}$, at $70°$ to the horizontal. This is represented by the arrow $\overrightarrow{OA}$ in Fig. 1.1, on a scale of $1\,\mathrm{cm}$ to $5\,\mathrm{m}$. To this must be added a displacement of magnitude $\frac{1}{2} \times 10 \times 1.5^2$ m, that is 11.25 m, vertically downwards, represented by the arrow $\overrightarrow{AB}$. The sum of these is the displacement $\overrightarrow{OB}$. So after 1.5 seconds the ball is at B. You could if you wish calculate the coordinates of B, or the distance OB, but in this example these are not asked for.

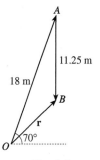

Fig. 1.1

EXAMPLE 1.1.2

A stone is thrown from the edge of a cliff with speed $18 \, \text{m s}^{-1}$. Draw diagrams to show the path of the stone in the next 4 seconds if it is thrown

a horizontally, **b** at $30°$ to the horizontal.

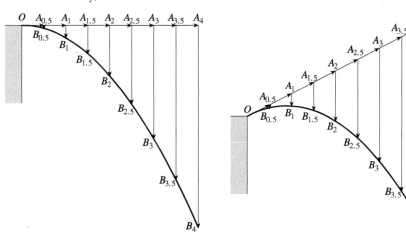

Fig. 1.2   Fig. 1.3

These diagrams were produced by superimposing several diagrams like Fig. 1.1. In Figs. 1.2 and 1.3 (for parts **a** and **b** respectively) this has been done at intervals of 0.5 s, that is for $t = 0.5, 1, 1.5, \ldots , 4$. The displacements $\mathbf{u}t$ in these times have magnitudes 9 m, 18 m, ... , 72 m. The vertical displacements have magnitudes 1.25 m, 5 m, 11.25 m, ..., 80 m. The points corresponding to A and B at time t are denoted by A_t and B_t.

You can now show the paths by drawing smooth curves through the points O, $B_{0.5}, B_1, \ldots, B_4$ for the two initial velocities.

The word **projectile** is often used to describe an object thrown in this way. The path of a projectile is called its **trajectory**.

A vector triangle can also be used to find the velocity of a projectile at a given time. If there were no gravity the velocity would have the constant value $\mathbf{u}$ indefinitely. The effect of gravity is to add to this a velocity of magnitude gt vertically downwards, which can be written as the vector $\mathbf{g}t$. This gives the equation

$$\mathbf{v} = \mathbf{u} + \mathbf{g}t,$$

assuming that air resistance can be neglected.

EXAMPLE 1.1.3

For the ball in Example 1.1.1, find the velocity after 1.5 seconds.

The vector **u** has magnitude $12\,\text{m s}^{-1}$ at $70°$ to the horizontal. The vector **g**t has magnitude $10 \times 1.5\,\text{m s}^{-1}$, that is $15\,\text{m s}^{-1}$, directed vertically downwards.

To draw a vector triangle you need to choose a scale in which velocities are represented by displacements. Fig. 1.4 is drawn on a scale of $1\,\text{cm}$ to $5\,\text{m s}^{-1}$. You can verify by measurement that the magnitude of **v** is about $5.5\,\text{m s}^{-1}$, and it is directed at about $42°$ below the horizontal.

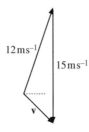

12 m s⁻¹   15 m s⁻¹   v

Fig. 1.4

Fig. 1.5 combines the results of Examples 1.1.1 and 1.1.3, showing both the position of the ball after 1.5 seconds and the direction in which it is moving.

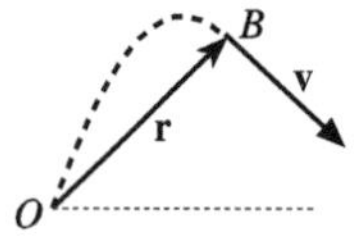

Fig. 1.5

As a reminder of what was said at the very start of this chapter, any object moving through the air will experience air resistance. How significant this is will depend upon a number of factors, including the nature of the object. For example, a feather is far more affected by air resistance than a solid metal ball. Also, experiments show that air resistance increases with speed. Air resistance is usually significant for objects moving at high speeds through large distances. In this chapter, it is assumed that air resistance is small enough that it can be ignored, and so all of the answers are only approximate.

Some people think that another limitation of this model for projectiles is the assumption that the acceleration is constant throughout the motion. They are aware that gravity reduces with height. However, while this last point is true, the acceleration change is miniscule: it only reduces by about 0.003% for every 100 metres above the ground. This is far less significant than any measurement error or air resistance effects for the type of projectiles we are considering, and so cannot be considered a significant limitation. The change in gravity does need to be taken into account when designing spacecraft launchers, but that is beyond the scope of this course.

Exercise 1A

1 A stone is thrown horizontally with speed $15\,\text{m s}^{-1}$ from the top of a cliff 30 metres high. Construct a diagram showing the positions of the particle at 0.5 second intervals. Estimate the distance of the stone from the thrower when it is level with the foot of the cliff, and the time that it takes to fall.

2 A pipe discharges water from the roof of a building, at a height of 60 metres above the ground. Initially the water moves with speed $1\,\mathrm{m\,s^{-1}}$, horizontally at right angles to the wall. Construct a diagram using intervals of 0.5 seconds to find the distance from the wall at which the water strikes the ground.

3 A particle is projected with speed $10\,\mathrm{m\,s^{-1}}$ at an angle of elevation of $40°$. Construct a diagram showing the position of the particle at intervals of 0.25 seconds for the first 1.5 seconds of its motion. Hence estimate the period of time for which the particle is higher than the point of projection.

4 A ball is thrown with speed $14\,\mathrm{m\,s^{-1}}$ at $35°$ above the horizontal. Draw diagrams to find the position and velocity of the ball 3 seconds later.

5 A particle is projected with speed $9\,\mathrm{m\,s^{-1}}$ at $40°$ to the horizontal. Calculate the time the particle takes to reach its maximum height, and find its speed at that instant.

6 A cannon fires a shot at $38°$ above the horizontal. The initial speed of the cannonball is $70\,\mathrm{m\,s^{-1}}$. Calculate the distance between the cannon and the point where the cannonball lands, given that the two positions are at the same horizontal level.

7 A particle projected at $40°$ to the horizontal reaches its greatest height after 3 seconds. Calculate the speed of projection.

8 A ball thrown with speed $18\,\mathrm{m\,s^{-1}}$ is again at its initial height 2.7 seconds after projection. Calculate the angle between the horizontal and the initial direction of motion of the ball.

9 A particle reaches its greatest height 2 seconds after projection, when it is travelling with speed $7\,\mathrm{m\,s^{-1}}$. Calculate the initial velocity of the particle. When it is again at the same level as the point of projection, how far has it travelled horizontally?

10 Two particles A and B are simultaneously projected from the same point on a horizontal plane. The initial velocity of A is $15\,\mathrm{m\,s^{-1}}$ at $25°$ to the horizontal, and the initial velocity of B is $15\,\mathrm{m\,s^{-1}}$ at $65°$ to the horizontal.

   a Construct a diagram showing the paths of both particles until they strike the horizontal plane.

   b From your diagram estimate the time that each particle is in the air.

   c Calculate these times, correct to 3 significant figures.

1.2 Coordinate methods

For the purposes of calculation it often helps to use coordinates, with column vectors representing displacements, velocities and accelerations, just as was done for forces in M1 Chapter 10. It is usual to take the x-axis horizontal and the y-axis vertical.

For instance, in Example 1.1.2(a), the initial velocity $\mathbf{u}$ of the stone was $18\,\mathrm{m\,s^{-1}}$ horizontally, which could be represented by the column vector $\begin{pmatrix} 18 \\ 0 \end{pmatrix}$. Since the units are metres and seconds, $\mathbf{g}$ is $10\,\mathrm{m\,s^{-2}}$ vertically downwards, represented by $\begin{pmatrix} 0 \\ -10 \end{pmatrix}$.

Denoting the displacement $\mathbf{r}$ by $\begin{pmatrix} x \\ y \end{pmatrix}$, the equation becomes $\mathbf{r} = \mathbf{u}t + \frac{1}{2}\mathbf{g}t^2$ becomes

$$\begin{pmatrix} x \\ y \end{pmatrix} = \begin{pmatrix} 18 \\ 0 \end{pmatrix} t + \frac{1}{2}\begin{pmatrix} 0 \\ -10 \end{pmatrix} t^2, \text{ or more simply } \begin{pmatrix} x \\ y \end{pmatrix} = \begin{pmatrix} 18t \\ 0 \end{pmatrix} + \begin{pmatrix} 0 \\ -5t^2 \end{pmatrix} = \begin{pmatrix} 18t \\ -5t^2 \end{pmatrix}.$$

You can then read off along each line to get the pair of equations

$$x = 18t \quad \text{and} \quad y = -5t^2.$$

From these you can calculate the coordinates of the stone after any time t.

You can make t the subject of the first equation as $t = \frac{1}{18}x$ and then substitute this in the second equation to get $y = -5\left(\frac{1}{18}x\right)^2$, or (approximately) $y = -0.015x^2$. This is the equation of the trajectory. You will recognise this as a parabola with its vertex at O.

You can do the same thing with the velocity equation $\mathbf{v} = \mathbf{u} + \mathbf{g}t$, which becomes

$$\mathbf{v} = \begin{pmatrix} 18 \\ 0 \end{pmatrix} + \begin{pmatrix} 0 \\ -10 \end{pmatrix} t = \begin{pmatrix} 18 \\ 0 \end{pmatrix} + \begin{pmatrix} 0 \\ -10t \end{pmatrix} = \begin{pmatrix} 18 \\ -10t \end{pmatrix}.$$

This shows that the velocity has components 18 and $-10t$ in the x- and y-directions respectively.

Notice that 18 is the derivative of $18t$ with respect to t, and $-10t$ is the derivative of $-5t^2$. This is a special case of a general rule.

> If the displacement of a projectile is $\begin{pmatrix} x \\ y \end{pmatrix}$, its velocity is $\begin{pmatrix} \dfrac{dx}{dt} \\ \dfrac{dy}{dt} \end{pmatrix}$.

This is a generalisation of the result given in M1 Section 11.2 for motion in a straight line.

Here is a good place to use the shorthand notation (dot notation) introduced in M1 Section 11.5, using $\dot{x}$ to stand for $\dfrac{dx}{dt}$ and $\dot{y}$ for $\dfrac{dy}{dt}$. You can then write the velocity vector as $\begin{pmatrix} \dot{x} \\ \dot{y} \end{pmatrix}$.

Now consider the general case, when the projectile starts with an initial speed u at an angle θ to the horizontal. Its initial velocity $\mathbf{u}$ can be described either in terms of u and θ, or in terms of its horizontal and vertical components p and q. These are connected by $p = u\cos\theta$ and $q = u\sin\theta$. The notation is illustrated in Figs. 1.6 and 1.7.

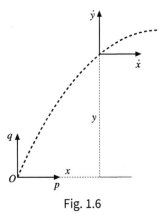

Fig. 1.6

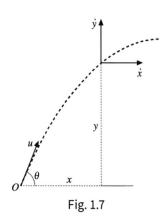

Fig. 1.7

The acceleration $\mathbf{g}$ is represented by $\begin{pmatrix} 0 \\ -g \end{pmatrix}$, so the equation

$$\mathbf{r} = \mathbf{u}t + \tfrac{1}{2}\mathbf{g}t^2$$

becomes

$$\begin{pmatrix} x \\ y \end{pmatrix} = \begin{pmatrix} pt \\ qt \end{pmatrix} + \begin{pmatrix} 0 \\ -\tfrac{1}{2}gt^2 \end{pmatrix} \quad \text{or} \quad \begin{pmatrix} x \\ y \end{pmatrix} = \begin{pmatrix} u\cos\theta\, t \\ u\sin\theta\, t \end{pmatrix} + \begin{pmatrix} 0 \\ -\tfrac{1}{2}gt^2 \end{pmatrix}.$$

By reading along each line in turn, the separate equations for the coordinates are

$$x = pt \qquad\qquad \text{or} \quad x = u\cos\theta\, t,$$
$$\text{and} \quad y = qt - \tfrac{1}{2}gt^2 \qquad \text{or} \quad y = u\sin\theta\, t - \tfrac{1}{2}gt^2.$$

In a similar way, $\mathbf{v} = \mathbf{u} + \mathbf{g}t$ becomes

$$\begin{pmatrix} \dot{x} \\ \dot{y} \end{pmatrix} = \begin{pmatrix} p \\ q \end{pmatrix} + \begin{pmatrix} 0 \\ -gt \end{pmatrix} \quad \text{or} \quad \begin{pmatrix} \dot{x} \\ \dot{y} \end{pmatrix} = \begin{pmatrix} u\cos\theta \\ u\sin\theta \end{pmatrix} + \begin{pmatrix} 0 \\ -gt \end{pmatrix}.$$

So

$$\dot{x} = p \qquad\qquad \text{or} \quad \dot{x} = u\cos\theta,$$
$$\text{and} \quad \dot{y} = q - gt \qquad \text{or} \quad \dot{y} = u\sin\theta - gt.$$

Since g, p, q, u and θ are all constant, you can see again that $\dot{x}$ and $\dot{y}$ are the derivatives of x and y with respect to t.

Now the equations $x = pt$ and $\dot{x} = p$ are just the same as those you would use for a particle moving in a straight line with constant velocity p. And the equations $y = qt - \tfrac{1}{2}gt^2$ and $\dot{y} = q - gt$ are the same as those for a particle moving in a vertical line with initial velocity q and acceleration $-g$. This establishes the **independence of horizontal and vertical motion**.

> If a projectile is launched from O with an initial velocity having horizontal and vertical components p and q, under the action of the force of gravity alone and neglecting air resistance, and if its coordinates at a later time are (x,y), then
>
> - the value of x is the same as for a particle moving in a horizontal line with constant velocity p;
>
> - the value of y is the same as for a particle moving in a vertical line with initial velocity q and acceleration $-g$.

EXAMPLE 1.2.1

A golf ball is driven with a speed of $45\,\text{m s}^{-1}$ at $37°$ to the horizontal across horizontal ground. How high above the ground does it rise, and how far away from the starting point does it first land?

> To a good enough approximation $\cos 37° = 0.8$ and $\sin 37° = 0.6$, so the horizontal and vertical components of the initial velocity are $p = 45 \times 0.8\,\text{m s}^{-1} = 36\,\text{m s}^{-1}$ and $q = 45 \times 0.6\,\text{m s}^{-1} = 27\,\text{m s}^{-1}$. The approximate value of g is $10\,\text{m}^{-2}$.

To find the height you only need to consider the y-coordinate. To adapt the equation $v^2 = u^2 + 2as$ with the notation of Fig. 1.6, you have to insert the numerical values u (that is q) $= 27$ and $a = -10$, and replace s by y and v by $\dot{y}$. This gives

$$\dot{y}^2 = 27^2 - 2 \times 10 \times y = 729 - 20y$$

When the ball is at its greatest height, $\dot{y} = 0$, so $729 - 20y = 0$. This gives $y = \frac{729}{20} = 36.45$.

To find how far away the ball lands you need to use both coordinates, and the link between these is the time t. So use the y-equation to find how long the ball is in the air, and then use the x-equation to find how far it goes horizontally in that time.

Adapting the equation $s = ut + \frac{1}{2}at^2$ for the vertical motion,

$$y = 27t - 5t^2.$$

When the ball hits the ground $y = 0$, so that $t = \frac{27}{5} = 5.4$. A particle moving horizontally with constant speed $36\,\mathrm{m\,s^{-1}}$ would go $36 \times 5.4\,\mathrm{m}$, that is $194.4\,\mathrm{m}$, in this time.

So, according to the gravity model, the ball would rise to a height of about 36 metres, and first land about 194 metres from the starting point.

In practice, these answers would need to be modified to take account of air resistance and the aerodynamic lift on the ball.

EXAMPLE 1.2.2

In a game of tennis a player serves the ball horizontally from a height of 2 metres. It has to satisfy two conditions.

i It must pass over the net, which is 0.9 metres high at a distance of 12 metres from the server.

ii It must hit the ground less than 18 metres from the server.

At what speeds can it be hit?

It is simplest to take the origin at ground level, rather than at the point from which the ball is served, so add 2 to the y-coordinate given by the general formula. If the initial speed of the ball is $p\,\mathrm{m\,s^{-1}}$,

$$x = pt \quad \text{and} \quad y = 2 - 5t^2.$$

Both conditions involve both the x- and y-coordinates, and the time t is used as the link.

i The ball passes over the net when $12 = pt$, that is $t = \dfrac{12}{p}$. The value of y is then $2 - 5\left(\dfrac{12}{p}\right)^2 = 2 - \dfrac{720}{p^2}$, and this must be more than 0.9. So

$$2 - \frac{720}{p^2} > 0.9.$$

This gives $\dfrac{720}{p^2} < 1.1$, which is $p > \sqrt{\dfrac{720}{1.1}} \approx 25.6$.

ii The ball lands when $y = 0$, that is when $2 - 5t^2 = 0$, or $t\sqrt{\dfrac{2}{5}}$. It has then gone a horizontal distance of $p\sqrt{\dfrac{2}{5}}$ metres, and to satisfy the second condition you need $p\sqrt{\dfrac{2}{5}} < 18$. This gives $p < 18\sqrt{\dfrac{5}{2}} \approx 28.5$.

So the ball can be hit with any speed between about $25.6\,\text{m s}^{-1}$ and $28.5\,\text{m s}^{-1}$.

EXAMPLE 1.2.3

A cricketer scores a six by hitting the ball at an angle of $30°$ to the horizontal. The ball passes over the boundary 90 metres away at a height of 5 metres above the ground. Neglecting air resistance, find the speed with which the ball was hit.

If the initial speed was $u\,\text{m s}^{-1}$, the equations of horizontal and vertical motion are

$$x = u\cos 30° t \quad \text{and} \quad y = u\sin 30° t - 5t^2.$$

You know that, when the ball passes over the boundary, $x = 90$ and $y = 5$. Using the values $\cos 30° = \tfrac{1}{2}\sqrt{3}$ and $\sin 30° = \tfrac{1}{2}$,

$$90 = u \times \tfrac{1}{2}\sqrt{3} \times t = \tfrac{1}{2}\sqrt{3}ut \quad \text{and} \quad 5 = u \times \tfrac{1}{2} \times t - 5t^2 = \tfrac{1}{2}ut - 5t^2$$

for the same value of t.

From the first equation, $ut = \dfrac{180}{\sqrt{3}} = 60\sqrt{3}$. Substituting this in the second equation gives $5 = 30\sqrt{3} - 5t^2$, which gives $t = \sqrt{6\sqrt{3} - 1} = 3.06\ldots$.

It follows that $u = \dfrac{60\sqrt{3}}{t} = \dfrac{60\sqrt{3}}{3.06\ldots} \approx 33.9$.

The initial speed of the ball was about $34\,\text{m s}^{-1}$.

EXAMPLE 1.2.4

A boy uses a catapult to send a small ball through his friend's open window. The window is 8 metres up a wall 12 metres away from the boy. The ball enters the window descending at an angle of $45°$ to the horizontal. Find the initial velocity of the ball.

One of the modelling assumptions we are making here is that the window is just a point. In the real world, the window has a significant height, so in a more sophisticated model, we could take this into account to obtain a range of possible velocities.

Denote the horizontal and vertical components of the initial velocity by $p\,\text{m s}^{-1}$ and $q\,\text{m s}^{-1}$. If the ball enters the window after t seconds,

$$12 = pt \quad \text{and} \quad 8 = qt - 5t^2.$$

Also, as the ball enters the window, its velocity has components $\dot{x} = p$ and $\dot{y} = q - 10t$. Since this is at an angle of $45°$ below the horizontal, $\dot{y} = -\dot{x}$, so $q - 10t = -p$, or

$$p + q = 10t.$$

You now have three equations involving p, q and t. From the first two

equations, $p = \dfrac{12}{t}$ and $q = \dfrac{8+5t^2}{t}$. Substituting these expressions in the third

equation gives $\dfrac{12}{t} + \dfrac{8+5t^2}{t} = 10t,$ that is

$$12 + (8 + 5t^2) = 10t^2, \quad \text{which simplifies to} \quad 5t^2 = 20.$$

So $t = 2$, from which you get $p = \dfrac{12}{2} = 6$

and $q = \dfrac{8+5\times 2^2}{2} = 14.$

Fig. 1.8 shows how these components are combined by the triangle
rule to give the initial velocity of the ball. This has magnitude

$\sqrt{6^2 + 14^2}\,\mathrm{m\,s^{-1}} \approx 15.2\,\mathrm{m\,s^{-1}}$ at an angle $\tan^{-1}\dfrac{14}{6} \approx 66.8°$ to the horizontal.

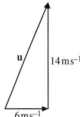

$14\,\mathrm{m\,s^{-1}}$

$6\,\mathrm{m\,s^{-1}}$

Fig. 1.8

The ball is projected at just over $15\,\mathrm{m\,s^{-1}}$ at $67°$ to the horizontal.

Exercise 1B

Assume that all motion takes place above a horizontal plane.

1   A particle is projected horizontally with speed $13\,\mathrm{m\,s^{-1}}$, from a point high above a
    horizontal plane. Find the horizontal and vertical components of the velocity of
    the particle after 2 seconds.

2   The time of flight of an arrow fired with initial speed $30\,\mathrm{m\,s^{-1}}$ horizontally from
    the top of a tower was 2.4 seconds. Calculate the horizontal distance from the
    tower to the arrow's landing point. Calculate also the height of the tower.

3   Show that the arrow in Question 2 enters the ground with a speed of about
    $38\,\mathrm{m\,s^{-1}}$ at an angle of about $39°$ to the horizontal.

4   A stone is thrown from the point O on top of a cliff with velocity $\begin{pmatrix} 15 \\ 0 \end{pmatrix}\mathrm{m\,s^{-1}}$. Find
    the position vector of the stone after 2 seconds.

5   A particle is projected with speed $35\,\mathrm{m\,s^{-1}}$ at an angle of $40°$ above the horizontal.
    Calculate the horizontal and vertical components of the displacement of the
    particle after 3 seconds. Calculate also the horizontal and vertical components of
    the velocity of the particle at this instant.

6   A famine relief aircraft, flying over horizontal ground at a height of 245 metres,
    drops a sack of food.

    a   Calculate the time that the sack takes to fall.

    b   Calculate the vertical component of the velocity with which the sack hits the
        ground.

c If the speed of the aircraft is $70\,\mathrm{m\,s^{-1}}$, at what distance before the target zone should the sack be released?

7 A girl stands at the water's edge and throws a flat stone horizontally from a height of $80\,\mathrm{cm}$.

   a Calculate the time the stone is in the air before it hits the water.

   b Find the vertical component of the velocity with which the stone hits the water.

     The girl hopes to get the stone to bounce off the water surface. To do this the stone must hit the water at an angle to the horizontal of $15°$ or less.

   c What is the least speed with which she can throw the stone to achieve this?

   d If she throws the stone at this speed, how far away will the stone hit the water?

8 A batsman tries to hit a six, but the ball is caught by a fielder on the boundary. The ball is in the air for 3 seconds, and the fielder is 60 metres from the bat. The batsman's hit and the fielder's catch are at the same height above the ground. Calculate

   a the horizontal component,

   b the vertical component of the velocity with which the ball is hit.

   Hence find the magnitude and direction of this velocity.

9 A stone thrown with speed $17\,\mathrm{m\,s^{-1}}$ reaches a greatest height of 5 metres. Calculate the angle of projection.

10 A particle projected at $30°$ to the horizontal rises to a height of 10 metres. Calculate the initial speed of the particle, and its least speed during the flight.

11 In the first 2 seconds of motion a projectile rises 5 metres and travels a horizontal distance of 30 metres. Calculate its initial speed.

12 The nozzle of a fountain projects a jet of water with speed $12\,\mathrm{m\,s^{-1}}$ at $70°$ to the horizontal. The water is caught in a cup 5 metres above the level of the nozzle. Calculate the time taken by the water to reach the cup.

13 A stone was thrown with speed $15\,\mathrm{m\,s^{-1}}$, at an angle of $40°$. It broke a small window 1.2 seconds after being thrown. Calculate the distance of the window from the point at which the stone was thrown.

14 A football, kicked from ground level, enters the goal after 2 seconds with velocity $\begin{pmatrix} 12 \\ -9 \end{pmatrix}\mathrm{m\,s^{-1}}$. Neglecting air resistance, calculate

   a the speed and angle at which the ball was kicked,

   b the height of the ball as it enters the goal,

   c the greatest height of the ball above the ground.

15 A particle is projected from a point O at an angle of $35°$ above the horizontal. At time Ts later the particle passes through a point A whose horizontal and vertically upward displacements from O are $8\,\mathrm{m}$ and $3\,\mathrm{m}$ respectively.

   i By using the equation of the particle's trajectory, or otherwise, find (in either order) the speed of projection of the particle from O and the value of T.

   ii Find the angle between the direction of motion of the particle at A and the horizontal.

(Cambridge International AS and A level Mathematics 9709/05 Paper 5 Q6 November 2007)

16 A tennis player hits the ball towards the net with velocity $\begin{pmatrix} 15 \\ 5 \end{pmatrix} \mathrm{ms}^{-1}$ from a point 8 metres from the net and 0.4 metres above the ground. The ball is in the air for 0.8 seconds before hitting the opponent's racket. Find, at the instant of impact with the opponent's racket,

a the velocity of the ball,

b the distance of the ball from the net and its height above the ground.

17 A projectile reaches its greatest height after 2 seconds, when it is 35 metres from its point of projection. Determine the initial velocity.

18 Particles A and B are projected simultaneously from the top T of a vertical tower, and move in the same vertical plane. T is 7.2 m above horizontal ground. A is projected horizontally with speed $8\,\mathrm{ms}^{-1}$ and B is projected at an angle of $60°$ above the horizontal with speed $5\,\mathrm{ms}^{-1}$. A and B move away from each other (see diagram).

i Find the time taken for A to reach the ground.

At the instant when A hits the ground,

ii show that B is approximately 5.2 m above the ground,

iii find the distance AB.

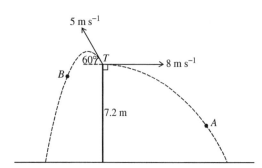

(Cambridge International AS and A level Mathematics 9709/05 Paper 5
Q5 June 2008)

19 A small ball B is projected from a point O with speed $14\,\mathrm{ms}^{-1}$ at an angle of $60°$ above the horizontal.

i Calculate the speed and direction of motion of B for the instant 1.8 s after projection.

The point O is 2 m above a horizontal plane.

ii Calculate the time after projection when B reaches the plane.

(Cambridge International AS and A level Mathematics 9709/51 Paper 5
Q4 November 2013)

20 A ski-jumper takes off from the ramp travelling at an angle of $10°$ below the horizontal with speed 72 kilometres per hour. Before landing she travels a horizontal distance of 70 metres. Find the time she is in the air, and the vertical distance she falls.

1.3 Some general formulae

Some of the results in this section use advanced trigonometry from P2&3 Chapter 5.

When you have more complicated problems to solve, it is useful to know formulae for some of the standard properties of trajectories. These are given in the notation of Fig. 1.7, which is repeated here as Fig. 1.9.

The formulae are based on the assumption that O is at ground level. If not, adjustments will be needed to allow for this.

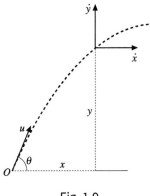

Fig. 1.9

i Greatest height

This depends only on the vertical motion of the projectile, for which the component of the initial velocity is $u \sin \theta$. The greatest height is reached when the vertical component of velocity is 0. By the usual constant acceleration formula, if h is the greatest height,

$$0^2 = (u \sin \theta)^2 - 2gh.$$

Therefore

$$h = \frac{u^2 \sin^2 \theta}{2g}.$$

ii Range on horizontal ground

If the ground is horizontal, the time at which the projectile lands is given by putting $y = 0$ in the equation $y = u \sin \theta t - \frac{1}{2}gt^2$. This gives $t = \dfrac{2u \sin \theta}{g}$. This is called the **time of flight**. If at this time the x-coordinate is r, then

$$r = u \cos \theta t = \frac{u \cos \theta \times 2u \sin \theta}{g} = \frac{2u^2 \sin \theta \cos \theta}{g}.$$

It is shown in P2&3 Section 5.5 that $2 \sin \theta \cos \theta$ is the expanded form of $\sin 2\theta$. So you can write the formula more simply as

$$r = \frac{u^2 \sin 2\theta}{g}.$$

iii Maximum range on horizontal ground

Suppose that the initial speed u is known, but that θ can vary. You will see from the graph of $\sin 2\theta$ (Fig. 1.10) that r takes its greatest value when $\theta = 45°$, and that $r_{max} = \dfrac{u^2}{g}$.

13

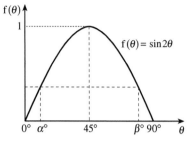

Fig. 1.10

Also, from the symmetry of the graph it follows that r has the same value when $\theta = \alpha°$ and when $\theta = (90 - \alpha)°$. So any point closer than the maximum range can be reached by either of two trajectories, one with a low angle of projection ($\alpha < 45$) and one with a high angle ($\beta = 90 - \alpha > 45$).

(iv) Equation of the trajectory

You can think of the equations for x and y in terms of t given in Section 1.2, above, as parametric equations for the trajectory, using time as the parameter. The cartesian equation can be found by turning $x = u\cos\theta\, t$ round to give $t = \dfrac{x}{u\cos\theta}$, and then substituting for t in the equation for y:

$$y = u\sin\theta \times \frac{x}{u\cos\theta} - \frac{1}{2}g\left(\frac{x}{u\cos\theta}\right)^2.$$

You can write this more simply by replacing $\dfrac{\sin\theta}{\cos\theta}$ by $\tan\theta$. Then

$$y = x\tan\theta - \frac{gx^2}{2u^2\cos^2\theta}.$$

Notice that, since u, g and θ are constant, this equation has the form $y = ax - bx^2$. You know that this is a parabola, and it is not difficult to show by differentiation that its vertex has coordinates $\left(\dfrac{a}{2b}, \dfrac{a^2}{4b}\right)$.

Writing $a = \tan\theta$ and $b = \dfrac{g}{2u^2\cos^2\theta}$, the vertex becomes $\left(\dfrac{u^2\sin 2\theta}{2g}, \dfrac{u^2\sin^2\theta}{2g}\right)$. This is another way of finding the formulae for the range and the greatest height.

Check the details for yourself.

For a projectile having initial velocity of magnitude u at an angle θ to the horizontal, under gravity but neglecting air resistance:

the greatest height reached is $\dfrac{u^2\sin^2\theta}{2g}$;

the time to return to its original height is $\dfrac{2u\sin\theta}{g}$;

the range on horizontal ground is $\dfrac{u^2\sin^2\theta}{g}$;

the maximum range on horizontal ground is $\dfrac{u^2}{g}$;

the equation of the trajectory is

$$y = x\tan\theta - \frac{gx^2}{2u^2\cos^2\theta} \quad \text{or} \quad y = x\tan\theta - \frac{gx^2\sec^2\theta}{2u^2}.$$

EXAMPLE 1.3.1

A basketball player throws the ball into the net, which is 3 metres horizontally from and 1 metre above the player's hands. The ball is thrown at $50°$ to the horizontal. How fast is it thrown?

Taking the player's hands as origin, you are given that $y = 1$ when $x = 3$ and that $\theta = 50°$. If you substitute these numbers into the equation of the trajectory you get

$$1 = 3\tan 50° - \frac{10 \times 9}{2u^2 \cos^2 50°}.$$

This gives

$$\frac{45}{u^2 \cos^2 50°} = 3\tan 50° - 1 = 2.575\ldots,$$

$$u^2 = \frac{45}{\cos^2 50° \times 2.575\ldots} = 42.29\ldots,$$

$$u = 6.50, \text{to 3 significant figures.}$$

The ball is thrown with a speed of about $6.5\,\mathrm{m\,s^{-1}}$.

EXAMPLE 1.3.2

A boy is standing on the beach and his sister is at the top of a cliff 6 metres away at a height of 3 metres. He throws her an apple with a speed of $10\,\mathrm{m\,s^{-1}}$. In what direction should he throw it?

You are given that $y = 3$ when $x = 6$ and that $u = 10$. It is more convenient to use the second form of the equation of the trajectory. Substituting the given numbers,

$$3 = 6\tan\theta - \frac{10 \times 6^2 \times \sec^2\theta}{2 \times 10^2},$$

which simplifies to

$$3\sec^2\theta - 10\tan\theta + 5 = 0.$$

To solve this equation you can use the identity $\sec^2\theta \equiv 1 + \tan^2\theta$. Then

$$3\left(1 + \tan^2\theta\right) - 10\tan\theta + 5 = 0, \qquad \text{that is} \qquad 3\tan^2\theta - 10\tan\theta + 8 = 0.$$

This is a quadratic equation for $\tan\theta$, which factorises as

$$(3\tan\theta - 4)(\tan\theta - 2) = 0, \qquad \text{so} \qquad \tan\theta = \tfrac{4}{3} \text{ or } \tan\theta = 2.$$

The apple should be thrown at either $\tan^{-1}\tfrac{4}{3}$ or $\tan^{-1} 2$ to the horizontal, that is either $53.1°$ or $63.4°$.

1.4 Accessible points

If you launch a projectile from O with a given initial speed u, but in an unspecified direction, you can obviously reach the points close to O, but not all points further away. You can use the method in Example 1.3.2 to find which points can be reached.

In Example 1.3.2 numerical values were given for x, y and u. If instead you keep these in algebraic form, then the equation of the trajectory can be written as

$$y = x\tan\theta - \frac{gx^2\left(1+\tan^2\theta\right)}{2u^2}.$$

This can then be arranged as a quadratic equation for $\tan\theta$,

$$gx^2\tan^2\theta - 2u^2x\tan\theta + \left(gx^2+2u^2y\right) = 0.$$

Now this equation can be solved to give values for $\tan\theta$ provided that the discriminant (that is, b^2-4ac in the usual notation for quadratics) is greater than or equal to 0. For this equation, the condition is

$$4u^4x^2 - 4gx^2\left(gx^2+2u^2y\right) \geq 0.$$

After cancelling $4x^2$, this can be rearranged as

$$y \leq \frac{u^2}{2g} - \frac{gx^2}{2u^2}.$$

Suppose, for example, that the initial speed is $10\,\mathrm{m\,s^{-1}}$. Then, in metre units, with $g=10$, $\dfrac{u^2}{g}=10$, so this inequality becomes

$$y \leq 5 - \tfrac{1}{20}x^2.$$

This is illustrated in Fig. 1.11, which shows several possible trajectories with this initial speed for various angles θ. All the points on these curves lie on or underneath the parabola with equation $y = 5 - \frac{1}{20}x^2$. This is called the **bounding parabola** for this initial speed. It separates the points which can be reached from O from those which can't.

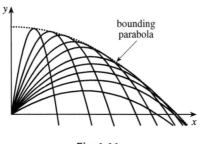

bounding parabola

Fig. 1.11

Thus in Example 1.3.2 the coordinates of the girl were (6,3). Since $5 - \frac{1}{20}\times 6^2 = 3.2 > 3$, these coordinates satisfy the inequality, so it is possible for her brother to throw the apple to reach her.

Exercise 1C

Assume that all motion takes place above horizontal ground unless otherwise stated.

1 A golfer strikes the ball with speed $60\,\mathrm{m\,s^{-1}}$. The ball lands in a bunker at the same level 210 metres away. Calculate the possible angles of projection.

2 A projectile is launched at $45°$ to the horizontal. It lands $1.28\,\mathrm{km}$ from the point of projection. Calculate the initial speed.

3 A footballer taking a free kick projects the ball with a speed of $20\,\mathrm{m\,s^{-1}}$ at $40°$ to the horizontal. Calculate the time of flight of the ball. How far from the point of the free kick would the ball hit the ground?

4 A stone being skimmed across the surface of a lake makes an angle of $15°$ with the horizontal as it leaves the surface of the water, and remains in the air for 0.6 seconds before its next bounce. Calculate the speed of the stone when it leaves the surface of the lake.

5 A projectile launched with speed $75\,\mathrm{m\,s^{-1}}$ is in the air for 14 seconds. Calculate the angle of projection.

6 An astronaut who can drive a golf ball a maximum distance of 350 metres on Earth can drive it 430 metres on planet Zog. Calculate the acceleration due to gravity on Zog.

7 An archer releases an arrow with speed $70\,\mathrm{m\,s^{-1}}$ at an angle of $25°$ to the horizontal. Calculate the range of the arrow. Determine the height of the arrow above its initial level when it has travelled a horizontal distance of 50 metres, and find the other horizontal distance for which it has the same height.

8 A particle projected at an angle of $40°$ passes through the point with coordinates (70,28) metres. Find the initial speed of the particle.

9 A hockey player taking a free hit projects the ball with speed $12.5\,\mathrm{m\,s^{-1}}$. A player 10 metres away intercepts the ball at a height of 1.8 metres. Calculate the angle of projection.

10 The equation of the path of a projectile is $y = 0.5x - 0.02x^2$. Determine the initial speed of the projectile.

11 A tennis player strikes the ball at a height of 0.5 metres. It passes above her opponent 10 metres away at a height of 4 metres, and lands 20 metres from the first player, who has not moved since striking the ball. Calculate the angle of projection of the ball.

12 In a game of cricket, a batsman strikes the ball at a height of 1 metre. It passes over a fielder 7 metres from the bat at a height of 3 metres, and hits the ground 60 metres from the bat. How fast was the ball hit?

13 The greatest height reached by a projectile is one-tenth of its range on horizontal ground. Calculate the angle of projection.

Miscellaneous exercise 1

1 A stone is projected horizontally with speed $8\,\mathrm{m\,s^{-1}}$ from a point O at the top of a vertical cliff. The horizontal and vertically upward displacements of the stone from O are $x\,\mathrm{m}$ and $y\,\mathrm{m}$ respectively.

   i Find the equation of the stone's trajectory.

   The stone enters the sea at a horizontal distance of 24m from the base of the cliff.

   ii Find the height above sea level of the top of the cliff.

   (Cambridge International AS and A level Mathematics 9709/05 Paper 5
   Q1 November 2006)

2 A particle is projected from horizontal ground with speed $u \, \text{ms}^{-1}$ at an angle of $\theta°$ above the horizontal. The greatest height reached by the particle is $10 \, \text{m}$ and the particle hits the ground at a distance of $40 \, \text{m}$ from the point of projection. In either order,

  i find the values of u and θ,

  ii find the equation of the trajectory, in the form $y = ax - bx^2$, where $x \, \text{m}$ and $y \, \text{m}$ are the horizontal and vertical displacements of the particle from the point of projection.

              (Cambridge International AS and A level Mathematics 9709/05 Paper 5
                                    Q4 November 2005)

3 A particle P is projected from a point O at an angle of $60°$ above horizontal ground. At an instant $0.6 \, \text{s}$ after projection, the angle of elevation of P from O is $45°$ (see diagram).

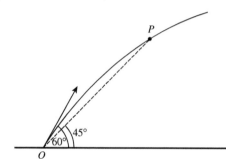

  i Show that the speed of projection of P is $8.20 \, \text{ms}^{-1}$, correct to 3 significant figures.

  ii Calculate the time after projection when the direction of motion of P is $45°$ above the horizontal.

              (Cambridge International AS and A level Mathematics 9709/51 Paper 5
                                    Q2 November 2011)

4 A particle is projected from a point O with speed $V \, \text{ms}^{-1}$ at an angle θ above the horizontal. After $0.3 \, \text{s}$ the particle is moving with speed $25 \, \text{ms}^{-1}$ at an angle $\tan^{-1}\left(\frac{7}{24}\right)$ above the horizontal.

  i Show that $V\cos\theta = 24$.

  ii Find the value of $V\sin\theta$ and hence find V and θ.

              (Cambridge International AS and A level Mathematics 9709/51 Paper 5
                                    Q4 November 2009)

5 A stone is projected from a point on horizontal ground with speed $25 \, \text{ms}^{-1}$ at an angle θ above the horizontal, where $\sin\theta = \frac{4}{5}$. At time $1.2 \, \text{s}$ after projection the stone passes through the point A. Subsequently the stone passes through the point B, which is at the same height above the ground as A. Find the horizontal distance AB.

              (Cambridge International AS and A level Mathematics 9709/05 Paper 5
                                    Q4 November 2006)

6 The equation of the trajectory of a small ball B projected from a fixed point O is

$$y = -0.05x^2,$$

where x and y are, respectively, the displacements in metres of B from O in the horizontal and vertically upwards directions.

  i Show that B is projected horizontally, and find its speed of projection.

  ii Find the value of y when the direction of motion is B is $60°$ below the horizontal, and find the corresponding speed of B.

              (Cambridge International AS and A level Mathematics 9709/51 Paper 5
                                    Q5 November 2014)

7 A fielder can throw a cricket ball faster at low angles than at high angles. This is modelled by assuming that, at an angle θ, he can throw a ball with a speed $k\sqrt{\cos\theta}$, where k is a constant.

 a Show that the horizontal distance he can throw is given by $\dfrac{2k^2}{g}\left(\sin\theta - \sin^3\theta\right)$.

 b Find the maximum distance he can throw the ball on level ground.

8 A particle P is projected with speed $20\,\text{m s}^{-1}$ at an angle of $40°$ above the horizontal from a point O on horizontal ground.

 i Find the height of P above the ground when P has speed $18\,\text{m s}^{-1}$.

 ii Calculate the length of time for which the speed of P is less than $18\,\text{m s}^{-1}$, and find the horizontal distance travelled by P during this time.

 (Cambridge International AS and A level Mathematics 9709/51 Paper 5 Q4 June 2014)

9 A particle is projected with speed $65\,\text{ms}^{-1}$ from a point on horizontal ground, in a direction making an angle of $\alpha°$ above the horizontal. The particle reaches the ground again after $12\,\text{s}$. Find

 i the value of α,

 ii the greatest height reached by the particle,

 iii the length of time for which the direction of motion of the particle is between $20°$ above the horizontal and $20°$ below the horizontal,

 iv the horizontal distance travelled by the particle in the time found in part iii.

 (Cambridge International AS and A level Mathematics 9709/05 Paper 5 Q9 June 2007)

10 A small ball B is projected with speed $15\,\text{m s}^{-1}$ at an angle of $41°$ above the horizontal from a point O which is $1.6\,\text{m}$ above horizontal ground. At time $t\,\text{s}$ after projection the horizontal and vertically upward displacements of B from O are $x\,\text{m}$ and $y\,\text{m}$ respectively.

 i Express x and y in terms of t and hence show that the equation of the trajectory of B is

 $y = 0.869x - 0.0390x^2$,

 where the coefficients are correct to 3 significant figures.

A vertical fence is $1.5\,\text{m}$ from O and perpendicular to the plane in which B moves. B just passes over the fence and subsequently strikes the ground at the point A.

 ii Calculate the height of the fence, and the distance from the fence to A.

 (Cambridge International AS and A level Mathematics 9709/51 Paper 5 Q7 June 2012)

19

Chapter 2
Moments

The objects in this chapter are not particles but rigid objects, which can turn about a point such as a hinge. When you have completed the chapter, you should

- understand the model of a rigid object
- understand that a rigid object has a centre of mass
- know that the turning effect of a force is measured by its moment
- be able to calculate moments of forces
- be able to solve problems about the equilibrium of rigid objects by taking moments about chosen points.

2.1 Rigid objects

You have probably used a spoon to open a tin of coffee. You slide the end of the handle into the gap between the rim of the tin and the lid, and prise the lid up by pushing down on the bowl of the spoon, as in Fig. 2.1.

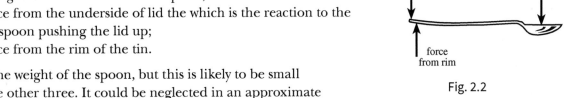

Fig. 2.1

There are three main forces acting on the spoon (Fig. 2.2):

- the force pushing down on the bowl of the spoon;
- the contact force from the underside of lid the which is the reaction to the force from the spoon pushing the lid up;
- the contact force from the rim of the tin.

Fig. 2.2

A fourth force is the weight of the spoon, but this is likely to be small compared with the other three. It could be neglected in an approximate calculation.

You couldn't use any old spoon to do this. The handle of a cheap metal spoon would probably bend when you push down on the bowl, and a plastic spoon would snap. You need a strong spoon which will keep its shape when you apply enough force to lift the lid.

An object which stays the same shape when forces are applied to it is said to be **rigid**.

The idea of a rigid object is another example of a mathematical model. In reality, it is impossible to make an object whose shape is completely unaffected by the forces applied to it. Geologists have shown that even the hardest rocks will compress and shear under the forces to which they are subjected by earth movements. But there are many objects whose behaviour approximates very closely to the rigid model.

Up to now you have modelled all the objects in mechanics problems as particles. This does not mean that they are necessarily small. A tree trunk, an aircraft, or even the earth can be treated as a particle so long as you are only interested in equilibrium or motion without rotation, with all the forces acting through the point where the particle is located.

But there is no way in which the spoon in Fig. 2.1 could be modelled as a particle. The whole point of using it is to apply a force acting along one line to produce a much larger force which acts along a different line. And when the lid starts to lift, the spoon will not move along a line but will rotate about the rim of the tin.

So the rigid object model is quite different from the particle model, and new principles and equations will be needed to solve problems about rigid objects.

2.2 Centres of mass

An important property of rigid objects is that they have a point at which, for many purposes, the whole mass may be supposed to be concentrated. This point is called the **centre of mass** of the object.

Take a rigid object with an irregular shape, such as a tennis racket. Lay it over the arm of a chair with the handle at right angles to the arm and the frame horizontal. Adjust the position of the racket on the arm until it balances, and then stick a piece of tape at the point of balance. The point on the central axis of the racket under the tape is the centre of mass.

Now fix a nail into a wall and hang the racket over the nail at some point of the string mesh. Whichever point you choose, you should find that the racket hangs with the centre of mass directly below the nail.

Go out to a field, hold the racket by its handle and throw it into the air. The racket will probably spin round quite erratically, but ignore this and keep your eye on the motion of the tape. This will move along quite a smooth trajectory, just like a ball thrown in a similar way. (If you have a video recorder, you can show this more precisely. Keep the camera still, and plot the path of the tape from a sequence of stills.)

Every rigid object has a centre of mass, but it can be quite complicated to calculate where it is. However, the centre of mass for objects with simple shapes is usually where you would expect it to be.

If an object is made of the same material with the same density all the way through, it is said to be **uniform**. For any uniform rigid object with a centre of geometrical symmetry, the centre of mass is at the centre of symmetry.

For example, the centre of mass of a uniform cylinder is at the mid-point of the axis; the centre of mass of a uniform cuboid is at the point where the diagonals meet; and so on.

Notice, though, that the centre of mass of an object may not actually be at a point of the physical object. For example, the centre of mass of a uniform hollow pipe lies on its axis, which is in open space.

2.3 The moment of a force

Take a strip of wood just over a metre long, the heavier the better so long as it is uniform. Mark a metre scale symmetrically on it, so that it can be used as a ruler. Drill a hole at the zero point of the scale (as in Fig. 2.3). Pin the ruler to a wall with a nail through the hole. (Make the diameter of the hole larger than that of the nail.)

Left alone, the ruler will hang vertically, its weight supported by the contact force at the nail.

Now support the ruler in a horizontal position by placing a finger underneath it at the mid-point of the scale, directly below the centre of mass, as in Fig. 2.3. The weight will then be completely supported by the contact force from your finger. If you look at the hole, you should find that the nail no longer exerts any force on the ruler.

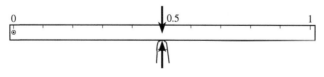

Fig. 2.3

Move your finger to the right end of the scale. You will now need to exert less force to support the ruler. If you look at the hole, you will see that the nail is in contact with the ruler at the top of the hole. The contact force from the nail is providing some of the force holding the ruler up, as in Fig. 2.4.

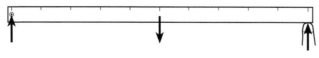

Fig. 2.4

Move your finger near to the left end of the scale. You will now have to exert a much larger force to keep the ruler horizontal. If you look at the hole, you will see that the nail is in contact with the ruler at the bottom of the hole. So you are having to push not only to support the weight of the ruler, but also against the contact force from the nail, as in Fig. 2.5.

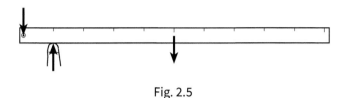

Fig. 2.5

To make this experiment more precise, you could replace your finger by a spring balance and measure the magnitude of the force. For a metre rule with a weight of 5 newtons, you might get a set of readings such as those in Table 2.6.

Table 2.6

Distance of support from nail (m)	0.5	1.0	0.1
Supporting force (N)	5	2.5	25

So if you double the distance of your finger from the nail (measured horizontally) from 0.5 m to 1 m, the force is halved from 5 N to 2.5 N. If you reduce the distance to one-fifth of its first value, from 0.5 m to 0.1 m, you multiply the force by five, from 5 N to 25 N.

The effect of a force varies according to the line along which it acts (see M1 Section 10.5). The experiment shows that if you multiply the magnitude of the supporting force by the distance of its line of action from the nail, you always get the same answer:

$$5 \times 0.5 = 2.5, \quad 2.5 \times 1.0 = 2.5, \quad \text{or} \quad 25 \times 0.1 = 2.5 .$$

The interpretation of this is that the product of the force and the distance is a measure of the turning effect of the force about the nail. This is called the 'moment' of the force about the nail.

> **The** moment **of a force about a point is calculated as the product of the magnitude of the force and the distance of its line of action from the point.**

In the SI system of units the moment of a force is measured in units of 'newton metres' (abbreviated to N m) because it is calculated by multiplying a number of newtons by a number of metres.

In the experiment with the ruler there is one force which stays the same, that is the weight of the ruler. This acts at a distance of 0.5 m from the nail and has magnitude 5 N, so its moment about the nail is 5×0.5 N m. The turning effect of the weight is clockwise, of magnitude 2.5 N m.

This has to be balanced by the turning effect of the force from your finger, which is a moment of 2.5 N m anticlockwise.

Suppose now that you place your finger under the ruler one-quarter of the way along, below the 0.25 m mark on the scale, and that the force from your finger is F newtons. Then the moment of this force is $F \times 0.25$ N m anticlockwise, so you can write an equation

$$F \times 0.25 = 2.5.$$

This equation is described as **taking moments** about the nail. A useful shorthand for showing where your equation comes from is to write

M (nail) $F \times 0.25 = 5 \times 0.5.$

That is, the anticlockwise moment of the force from your finger about the nail is equal to the clockwise moment of the weight. It follows that $F = 10$, so that in this position your finger must exert a force of 10 newtons.

Some people prefer to write equations of moments in the form

'anticlockwise moments – clockwise moments = 0'

rather than

'anticlockwise moments = clockwise moments.'

For example, the M (nail) equation on the previous page would be written as $F \times 0.25 - 5 \times 0.5 = 0$. If you go on to study the mechanics of rotation you will need to get used to putting all the moments on the left side of the equation. The drawback is the complication of having minus signs in the equation. Whichever form you use, it is very important to consider for each force whether its moment is anticlockwise or clockwise.

EXAMPLE 2.3.1

A door is kept closed by a wedge, placed under the door at a distance of 75 cm from the hinge. A person tries to open the door by applying a force of 40 N at 60 cm from the hinge, but it doesn't move. Calculate the friction force between the wedge and the floor.

Fig. 2.7

Fig. 2.7 shows the forces on the door which act horizontally. You must remember to put in the force from the hinge, even though it is not asked for.

There are also some vertical forces, the weight of the door and a vertical component of the hinge force. These have not been included in the diagram; they have no turning effect about the hinge.

It is simplest to use units of newton centimetres (N cm) for the moments. If the friction force is F newtons,

M (hinge) $F \times 75 = 40 \times 60.$

So $F = \dfrac{40 \times 60}{75} = 32$

The friction force between the wedge and the floor is 32 newtons.

If you prefer to stick with basic SI units, with forces in newtons and distances in metres, then the equation of moments would be $F \times 0.75 = 40 \times 0.6$. Using centimetres instead of metres avoids the need for decimals, so the arithmetic is simpler. In the next example the masses are very large, so it is simpler to work with mass in tonnes rather than kilograms, and force in kilonewtons rather than newtons, to avoid having too many zeros in the numbers. But whenever you are doing mechanics with numerical data, it is important to remember what units you are working in. If you use non-standard units, it helps to indicate this by writing a note about units alongside your solution.

EXAMPLE 2.3.2

Fig. 2.8 shows a simplified model of the horizontal arm of a crane which is being used to lift a load of 8 tonnes at a distance of 30 metres from the vertical column. The arm itself has a mass of 1.4 tonnes, distributed uniformly; its length is 50 metres, of which 10 metres extends on the opposite side of the column. Equilibrium is maintained by a counterbalance whose centre of mass is 9 metres from the column. Calculate the mass of the counterbalance.

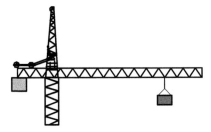

Fig. 2.8

In a numerical example like this, where you are given the masses of the components and have to find the weights, it is often a good idea to use the letter g to stand for the number 10. (This was discussed in M1 Section 7.5.) In this example you will find that g cancels out from the equations, so the answer is the same whatever value is taken for the acceleration due to gravity.

The forces on the arm, in kilonewtons, are shown in Fig. 2.9. Notice that, although the arm extends on both sides of the column, it is still possible to take its weight of $1.4g$ kN to be concentrated at the centre of mass, which is 15 metres to the right of the column. Let the mass of the counterbalance be M tonnes, so that its weight is Mg kN.

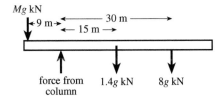

Fig. 2.9

Two of the forces, the weights of the arm and the load, have clockwise moments, and these are balanced by the anticlockwise moment of the weight of the counterbalance.

M (column) $\left(Mg\right) \times 9 = 1.4g \times 15 + 8g \times 30.$

This gives $M = \dfrac{(21 + 240)g}{9g} = 29.$

To maintain equilibrium, a counterbalance of mass 29 tonnes is needed.

Exercise 2A

1 A closed carpark barrier is modelled as a light rod which is pivoted at a point. The rod carries a weight of 200 N at a point 0.25 metres from the pivot, and it is supported horizontally at a point 1.8 metres from the pivot, on the other side from the weight. Find the magnitude of the vertical force exerted on the rod by the support.

2 A construction worker rides at one end of a uniform beam which is being moved into position in the skeleton of a high building, as shown in the diagram. The weight of the beam is 4200 N and the weight of the construction worker is 800 N. Given that the length of the beam is 12 m, find the distance between the construction worker and the crane's vertical chain when the beam is horizontal and the crane and the beam are at rest.

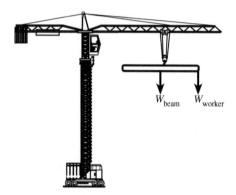

3 The total mass of two children is 60 kg. Find their separate masses, given that the children are balanced on a seesaw when one is 1.6 metres from the centre and the other is 1.4 metres from the centre.

   When the heavier child sits 1.6 metres from the centre and the lighter child sits 1.4 metres from the centre, the seesaw is balanced by applying a vertically downward force, behind the lighter child, at a distance of 2 metres from the centre. Find the magnitude of this downward force.

4 The diagram shows a door being pushed on one side with a force of magnitude 60 N, at a distance of 69 cm from the hinged edge, and on the other side by a force of magnitude 90 N. The forces act at the same height, and both act at right angles to the plane of the door. Given that the door does not move, find the distance from the hinged edge at which the force of magnitude 90 N acts.

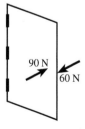

5 In each of the following cases a bone or collection of bones is modelled by a light rod. The rod is pivoted at P and carries a vertical load W newtons. The rod is held in position by the tension, acting vertically, in a tendon which connects the bone to a muscle. The magnitude of the tension is T newtons.

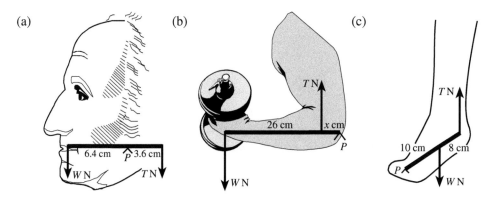

(a)   (b)   (c)

a   Find the value of T given that $W = 27$.

b   Given that $W = 40$ and $T = 300$, find x.

c   Given that $T = 200$, find W.

6   Two men, each of mass 80 kg, and a woman of mass 64 kg, are walking on a horizontal straight path which forms an edge of a lake. The woman's expensive hat is blown into the lake, and is caught by reeds at a distance of 4 metres from the path. A rigid plank, of length 6 metres and whose weight can be ignored, is available for use in retrieving the hat. The plank is placed on the path at right angles to the edge of the lake, and then pushed out so that it overhangs the lake. Two people stand on the part of the plank which is in contact with the path. The third person walks out along the plank over the lake. Determine whether

a   one of the men can retrieve the hat,

b   the woman can retrieve the hat.

7   Two equal uniform planks, A and B, of length 2 m are placed on a platform at right angles to the edge, as shown in the diagram. Keeping B still, A is pushed to the right as far as possible without tipping over. Then B is pushed to the right as far as possible. How far will the end of A project beyond the edge of the platform when B is just about to tip over?

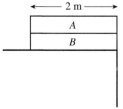

2.4   Forces from supports

In the experiment with the ruler in the last section, the supporting force from your finger was calculated from an equation of moments about the nail. But there are other ways of looking at it.

Fig. 2.5 shows the forces in the case when your finger is close to the nail, and it was found that at the hole the nail exerted a downward force on the ruler. You could calculate this force by imagining your finger as the hinge about which the ruler could turn. In this case the weight of the ruler, 5 N, is acting along a line 0.4 m from your finger, so that it has a clockwise moment of 5×0.4 N m, that is 2 N m.

This is balanced by the anticlockwise moment of the force from the nail, at a distance of 0.1 m from your finger. This force is therefore $\dfrac{2}{0.1}$ N, that is 20 N.

The principle of moments **If a rigid object is in equilibrium, the sum of the anticlockwise moments about any point must equal the sum of the clockwise moments.**

You could, of course, have calculated the force from the nail directly by resolving vertically for the forces on the ruler. It was shown earlier in Table 2.6 that the upward force from your finger is 25 N, and this must equal the sum of the weight of 5 N and the force from the nail. So the force from the nail is 20 N. This gives you a chance to check on your calculations.

EXAMPLE 2.4.1

A horizontal uniform plank of length 10 m and mass 50 kg is supported by two vertical ropes, attached at A, 2 m from the left end, and at B, 3 m from the right end. Two children stand on the plank: Lindi (44 kg) 4 m from the left end, and Jakob (36 kg) 2 m from the right end, as shown in Fig. 2.10. Find the tension in the ropes.

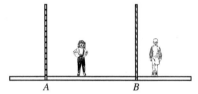

Fig. 2.10

There is no hinge here, but you can still take various points of the plank and consider the turning effects of the forces about them. The weights of the children and the plank are 440, 360 and 500 newtons. Let the tensions in the ropes at A and B be S and T newtons respectively, as shown in Fig. 2.11.

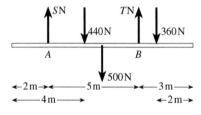

Fig. 2.11

For example, the tension T newtons has an anticlockwise moment about A, and this is balanced by the clockwise moments of the three weights.

$$M(A) \qquad T \times 5 = 440 \times 2 + 500 \times 3 + 360 \times 6,$$

which gives $T = 908$.

Now consider the turning effects about B. The weights of Lindi and the plank have anticlockwise moments, but the moment of Jakob's weight is clockwise.

$$M(B) \qquad 440 \times 3 + 500 \times 2 = 360 \times 1 + S \times 5,$$

which gives $S = 392$.

So the tensions in the ropes at A and B are 392 N and 908 N.

As a check, note that the sum of the two tensions is 1300 N, which is equal to the sum of the three weights.

In an example such as this, you are not restricted to taking moments about points at which the plank is attached to a rope. You could, for example, consider the moments about the left end of the plank. The three weights have a combined clockwise moment of $(440 \times 4 + 500 \times 5 + 360 \times 8)$ Nm, that is 7140 Nm, and the two tensions have a combined anticlockwise moment of $(S \times 2 + T \times 7)$ Nm. You can check that, with the values calculated, $S \times 2 + T \times 7 = 392 \times 2 + 908 \times 7 = 7140$.

You can in fact get a correct equation by taking moments about any point you like. The advantage of taking moments about A and B is that only one of the unknown forces comes into the equations. This is an example of a general strategy.

You can remove an unknown or unwanted force from an equation by taking moments about a point on its line of action.

EXAMPLE 2.4.2

A heavy rod AB of length 3 m lies on horizontal ground. To lift the end B off the ground needs a vertical force of 200 N. To lift A off the ground needs a force of 160 N. Find the weight of the rod, and the position of its centre of mass.

Let the weight of the rod be W newtons, and let the centre of mass be x metres from A. The two situations described are illustrated in Fig. 2.12.

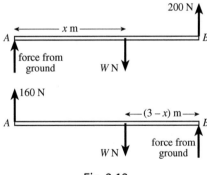

Fig. 2.12

When the end B is lifted:

$\mathsf{M}(A)$ $W \times x = 200 \times 3$.

When the end A is lifted:

$\mathsf{M}(B)$ $W \times (3 - x) = 160 \times 3$.

Adding these equations gives $W \times 3 = 1080$,

so that $W = 360$. Then, from the first equation, $x = \dfrac{600}{360} = \dfrac{5}{3}$.

The rod has a weight of 360 N, and its centre of mass is $1\frac{2}{3}$ m from A.

EXAMPLE 2.4.3

A car with front-wheel drive has mass 1000 kg, including the driver and passengers. The front wheels are 2.5 m in front of the rear wheels, and the centre of mass is 1.5 m in front of the rear wheels. The engine may be taken to have unlimited power. The coefficient of friction between the tyres and the road surface is 0.4.

a What is the maximum acceleration of which the car is capable?

b What difference would it make if bicycles of mass 75 kg were strapped to the back of the car 0.8 m behind the rear wheels?

The forward force on the car is produced by the friction at the front tyres. This cannot be more than 0.4 times the normal force S newtons between the front tyres and the ground. The normal force can be found by taking moments about the contact P between the rear tyres and the ground.

a The forces are shown in Fig. 2.13.

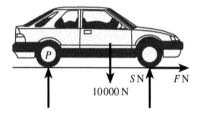

Fig. 2.13

$\mathcal{M}(P)$ $S \times 2.5 = 10000 \times 1.5$,

so $S = 6000$. The maximum forward force is therefore 0.4×6000 N, that is 2400 N. Therefore, by Newton's second law, the maximum acceleration is $\dfrac{2400}{1000}$ m s^{-2}, that is 2.4 m s^{-2}.

b There is now an additional anticlockwise moment of 750×0.8 N m.

$\mathcal{M}(P)$ $S \times 2.5 + 750 \times 0.8 = 10000 \times 1.5$,

which reduces the value of S to 5760 N. The maximum acceleration is then $0.4 \times \dfrac{5760}{1075}$ ms^{-2}, or about 2.14 m s^{-2}.

Exercise 2B

1 A cyclist sits astride his stationary machine, with his feet lightly touching the ground and his hands lightly touching the handlebar. The mass of the machine is 14 kg , and the line of action of its weight is the perpendicular bisector of $O_f O_r$, where O_f and O_r are the centres of the front and rear wheels respectively. The mass of the cyclist is 80 kg and his weight acts in a line through the saddle. Given that the distance $O_f O_r$ is 1020 mm and that the vertical line through the saddle is 240 mm in front of O_r, find the magnitude of the force exerted by the ground on

a   the front wheel,   **b**   the rear wheel.

2 A construction worker rides on a structural beam which is being moved into position in the skeleton of a high building, as shown in the diagram. The mass of the beam and the mass of the worker are 2000 kg and 80 kg respectively. The length of the beam is 12 m, and chains which hang vertically from the crane are attached at the left end of the beam and at a distance of 10 m from the left end. The worker sits at a distance of 7 m from the left end. Find the tension in each of the chains when the beam is horizontal and at rest.

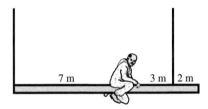

3 A uniform beam, of mass 16 kg and length 12 m , rests horizontally on supports at 2 m from its left end and 4 m from its right end. Find the force exerted on the beam by each of the supports, when a child of mass 32 kg and an adult of mass 70 kg stand on the beam at the left end and the right end respectively, as shown in the diagram.

Find also the mass which the adult would need to be carrying for the beam to be just on the point of tilting.

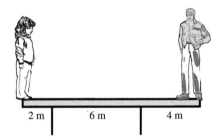

4 Two men A and B carry a boat horizontally over their heads along a shore. The length of the boat is 17.5 m. One end of the boat overhangs A's point of support by 2.5 m, and B's point of support is at the other end of the boat. A supports a load of 720 N and B supports a load of 480 N. Calculate the distance from B's end of the point through which the weight of the boat acts.

5 A uniform curtain rail of length 2.5 m and mass 5 kg rests horizontally on supports at its ends. Two curtains, each of mass 4 kg, hang from the rail. The left curtain is drawn across to the middle of the rail and the right curtain is drawn back to within 20 cm of the right support, as shown in the diagram. Assuming the weight of the left curtain acts at a distance of 0.625 m from the left end of the rail, and that the weight of the right curtain acts at a distance of 10 cm from the right end of the rail, find the magnitude of the vertical force exerted by each of the supports on the rail.

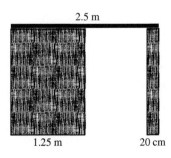

The right curtain is now drawn partially across the window with the effect that the magnitude of the vertical force exerted by the right support decreases by 4.8 N. Making a suitable assumption, which should be stated, find the distance from the right support to which the right curtain is now drawn.

6 A light rigid rod rests horizontally on supports at its ends. The rod carries two equal point loads, one at distance x from one end and the other at distance x from the other end. Show that the magnitude of the vertical force acting on the rod at each of the supports is independent of x.

A light clothes rail of length 2.5 m rests horizontally on supports at its ends. The rail has 25 identical garments hanging on it, at points distant $(10n-5)$ cm from one end, for $n = 1, 2, ..., 25$. The total weight of the garments is 250 N. State the magnitude of the vertical force acting on the rod at each of the supports.

The garments 5th and 14th from one end are removed. Find the magnitude of the vertical force now acting on the rod at each of the supports.

7 A cyclist and her machine have total mass 90 kg. The distance $O_f O_r$, where O_f and O_r are the centres of the front and rear wheels respectively, is 1 m. The total weight of the cyclist and her machine acts in a line 0.3 m in front of O_r.

    a Find the magnitude of the normal contact force exerted by the ground on the rear wheel.

The coefficient of friction between the rear wheel and the ground is 0.45.

    b Find the maximum possible frictional force exerted by the ground on the rear wheel.

    c Hence find the maximum possible acceleration of the cyclist and her machine.

2.5 Forces in different directions

The examples so far have all involved forces in parallel directions, but the principle of moments applies whatever the direction of the forces.

Imagine for example a rotating circular platform on a children's playground. You can turn this by exerting a horizontal force around the circumference. The turning effect of this force is the same, whatever way you are facing as you push.

EXAMPLE 2.5.1

A uniform rectangular plate of weight W is held in a vertical plane as shown in Fig. 2.14. The plate has width a and height b. It is hinged at the lower left corner, and kept in equilibrium by a horizontal force F applied at the upper right corner. Find F in terms of a, b and W.

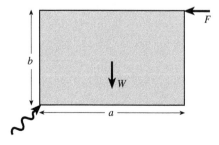

Fig. 2.14

When you show the forces in the diagram, it is important to include the force from the hinge as well as W and F. Draw it with a wavy line, as you do not know the exact direction in which it acts. For this reason take moments about the hinge, so this unknown force does not come into the equation.

The distances from the hinge to the lines of action of W and F are $\frac{1}{2}a$ and b respectively.

M (hinge) $\qquad Fb = W\left(\frac{1}{2}a\right).$

It follows directly that $F = \dfrac{Wa}{2b}.$

EXAMPLE 2.5.2

A bridge of weight W newtons is 4 metres long. It is supported by horizontal hinges along one edge, so that it can be raised to let boats pass underneath. It is raised by a cable attached to the opposite edge which passes over a pulley 8 metres above the hinge. Find the tension in the cable which will support the weight of the bridge

a when it is horizontal (Fig. 2.15),

b when the cable is perpendicular to the bridge (Fig. 2.16).

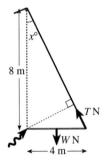

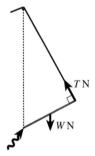

Fig. 2.15 Fig. 2.16

Denote the tension in the cable by T newtons. The force at the hinge is unknown, so T is calculated by taking moments about the hinge.

a The moment of the weight is $2W$ N m, but to find the moment of the tension needs trigonometry. Begin by noticing that the length of the cable up to the pulley is $4\sqrt{5}$ m.

So, if the cable makes an angle $x°$ with the vertical,

$$\cos x° = \frac{2}{\sqrt{5}} \quad \text{and} \quad \sin x° = \frac{1}{\sqrt{5}}$$

There are two ways of calculating the moment of the tension.

Method 1 The perpendicular distance from the hinge to the cable is $8\sin x°\,\text{m} = \dfrac{8}{\sqrt{5}}\,\text{m}$. So the moment of the tension is $\dfrac{8T}{\sqrt{5}}\,\text{N m}.$

33

Method 2   Split the tension into two components, $T\cos x°$N vertically and $T\sin x°$N horizontally, as in Fig. 2.17. The vertical component has moment

$$4T\cos x°\text{N m} = \frac{8T}{\sqrt{5}} = \text{N m}$$ The horizontal component has zero moment,

since its line of action passes through the hinge.

$$\mathsf{M}\text{(hinge)} \quad \frac{8T}{\sqrt{5}} = 2W.$$

Therefore $T = \dfrac{1}{4}\sqrt{5}W.$

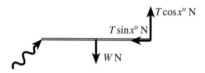

Fig. 2.17

b When the cable is perpendicular to the bridge, the angle between the bridge and the vertical is $\cos^{-1}\frac{4}{8} = 60°$. The perpendicular distance from the hinge to the cable is 4 metres, so the moment of the tension is $4T$ N m. To find the moment of the weight you need the perpendicular distance from the hinge to its line of action, which is $2\cos 30°$ metres, that is $\sqrt{3}$ metres.

$$\mathsf{M}\text{(hinge)} \qquad 4T = \sqrt{3}W.$$

Therefore $T = \frac{1}{4}\sqrt{3}W.$

EXAMPLE 2.5.3

A crate of weight W newtons is 1.8 m long and 0.8 m high. It is carried up a staircase at an angle of $10°$ to the horizontal by men supporting the crate by vertical forces at its two lower corners A and B. What proportion of the total load is carried by the lower man?

Since only the force from the lower man is asked for, take moments about the corner B held by the upper man. Let the supporting forces be P newtons and Q newtons, as shown in Fig. 2.18; the weight of the crate is W newtons.

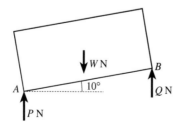

Fig. 2.18

Although all the forces are vertical, it is quite complicated to calculate the moments.

Method 1   The moments about B of the forces P newtons and the weight are Px N m and Wy N m, where x metres and y metres are the distances marked in Fig. 2.19.

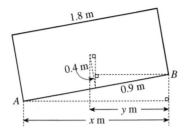

Fig. 2.19

These can be calculated from the right-angled triangles as

$$x = 1.8 \cos 10°$$

and $y = 0.9 \cos 10° + 0.4 \sin 10°.$

$\mathcal{M}(B)$ $W(0.9\cos 10° + 0.4\sin 10°) = P(1.8\cos 10°).$

Method 2   Split the forces into components parallel to the sides of the crate, as in Fig. 2.20. The force of $P\sin 10°$ N has no moment about B, since its line of action passes through B.

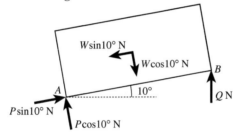

Fig. 2.20

$\mathcal{M}(B)$ $P\cos 10° \times 1.8 = W\cos 10° \times 0.9 + W\sin 10° \times 0.4.$

You can easily see that both equations are the same and give

$$P = \frac{W}{1.8} \times (0.9 + 0.4\tan 10°) = 0.539W \text{ , correct to 3 significant figures.}$$

The man supporting the crate at A must exert a force of about $0.54W$, which is 54% of the total load.

Exercise 2C

1   Find the moment of the force shown, about the point O, in each of the following cases.

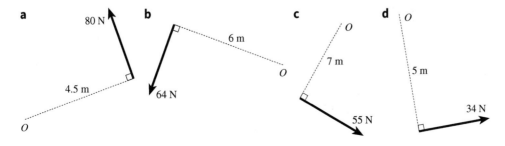

2 Find the total moment about O of the forces shown, in each of the following cases.

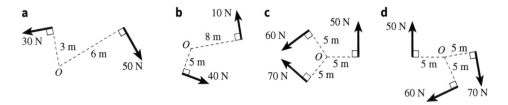

3 Find the moment of the force shown, about the point O, in each of the following cases.

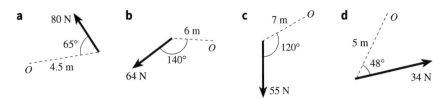

4 Find the value of d in each of the following cases, given that the moment about O is $480\,\text{N\,m}$ clockwise.

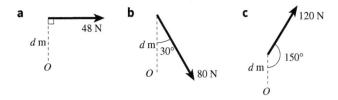

5 Find the value of θ in each of the following cases, given that the moment about O is $480\,\text{N\,m}$ anticlockwise.

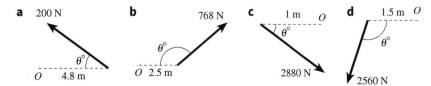

6 A capstan is used to wind a cable attached to a boat. It rotates round a vertical axis, and the radius of the drum on which the cable winds is $10\,\text{cm}$. Power is provided by six men pushing on the arms of the capstan at $90\,\text{cm}$ from the axis, each exerting a force of $200\,\text{N}$. Calculate the tension in the cable.

7 A mirror of mass $24\,\text{kg}$ is supported by two screws $80\,\text{cm}$ apart symmetrically placed along the top edge. One of the screws breaks, and while it is being replaced a helper keeps the mirror in position by a horizontal force applied 120 cm below the level of the screws. How large must this force be?

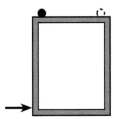

8 A uniform rod of length 2 metres and weight 40 newtons is hinged at one end. It is pulled aside at 50° to the vertical by a horizontal force applied at the other end. Find the magnitude of this force.

9 A hoop of weight 20 newtons can rotate freely about a pin fixed in a wall. A string has one end attached to the pin, runs round the circumference of the hoop to its lowest point, and is then held horizontally at its other end. A gradually increasing horizontal force is now applied to the string, so that the hoop begins to rotate about the pin. Find the tension in the string when the hoop has rotated through 40°.

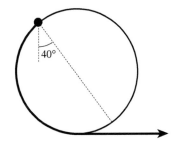

10 On a model ship, the mast OC has length 50 cm and weight 20 newtons. The mast is hinged to the deck at O, so that it can rotate in the vertical fore-and-aft plane of the ship. Small smooth rings are fixed at points A and B on the deck in this plane such that $AO = OB = 50$ cm. Threads from C are passed through these rings, and held at their ends by two children who exert forces of P newtons and Q newtons respectively. If $Q = 10$, calculate the value of P needed to hold the mast in equilibrium at 40° to the horizontal deck.

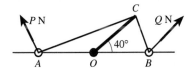

11 A chest, 2.4 metres long and 0.8 metres high, of weight 1500 newtons, stands on a rough floor. It is tilted about one edge of the base so that the length of the chest makes an angle of 8° with the horizontal. It is supported in this position by a prop at the mid-point of the parallel edge of the base. The prop is set perpendicular to the base of the chest, with its other end on the floor. Calculate the thrust in the prop.

12 In Greek myth Sisyphus was condemned for eternity to push a boulder up a hill. The boulder is modelled as a sphere of weight W and radius r. The hill slopes at an angle a to the horizontal, and Sisyphus exerts a force directed along a radius at an angle β to the hill. Prove that, to hold the boulder in equilibrium, he must exert a force of magnitude $\dfrac{W \sin \alpha}{\cos \beta}$.

13 A uniform rectangular hatch *ABCD*, of mass 12 kg, is hinged along its edge *AB*, and is horizontal when closed. A force acting through the mid-point of *CD* keeps the hatch open at a fixed angle of 25°.

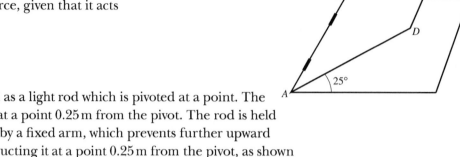

Find the magnitude of the force, given that it acts

  a vertically,

  b horizontally,

  c at right angles to *ABCD*.

14 A carpark barrier is modelled as a light rod which is pivoted at a point. The rod carries a weight of 200 N at a point 0.25 m from the pivot. The rod is held open at 80° to the horizontal by a fixed arm, which prevents further upward movement of the rod by obstructing it at a point 0.25 m from the pivot, as shown in the diagram. Find the force exerted on the rod by the fixed arm.

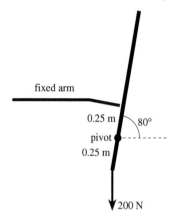

Find the magnitude of the force applied at a distance of 2 m from the pivot which will just start to move the rod downwards to its closed position, given that its direction is

  a vertical, **b** horizontal, **c** at right angles to the rod.

Miscellaneous exercise 2

1 A see-saw has mass 200 kg. It consists of a uniform beam of wood of length 3 m balanced on a pivot in the middle of the beam. Two boys are sitting on it as shown in the diagram.

Boy A has mass 50 kg and boy B has mass 40 kg. Boy B is sitting 10 cm from one end of the see-saw. How far is boy A from the other end, given that the see-saw is in equilibrium?

2 A uniform bench of length 2 m and mass 80 kg rests on two supports, *A* and *B*, as shown in the diagram. Support *A* is 40 cm from one end of the bench and support *B* is at the other end. Determine the magnitudes of the contact forces on the two supports.

3 A straight board *ABCD* is 10 metres long and rests on two supports, one at *B* and one at *C*, where *AB = CD = 2* m. The board has weight *W* N and the centre of mass of the board lies *x* m from *B*, as shown.

When a woman of mass 60 kg stands on the board 1 m from A, the board is on the point of toppling. When the same woman stands on the board 1.2 m from D, the board is also on the point of toppling. Calculate the values of W and x.

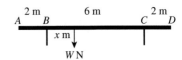

4 A uniform rod AB of mass 17 kg and length 0.8 m is attached to a wall with a smooth hinge at A and is supported in position by a light inextensible string of length 1.5 m. The string is attached to the end B of the rod and to a point C vertically above A. The string makes a right angle with the rod. Determine the tension in the string.

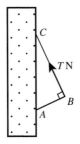

5 A uniform rod AB of weight WN is smoothly hinged to a vertical wall at A. It is held in equilibrium at $30°$ to the wall by means of a light inextensible string making an angle of $50°$ with the rod. The tension in the string is 100 N.

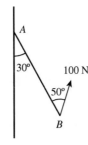

a Find W.

b Without performing any further calculations, explain briefly whether the tension in the string will be greater or less than 100 N if the string now makes an angle of $60°$ with the rod, with the rod remaining in equilibrium at $30°$ to the wall

6

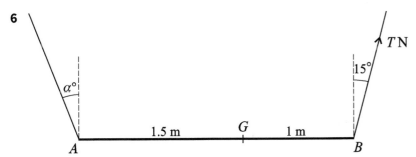

A non-uniform rod AB of length 2.5 m and mass 3 kg has its centre of mass at the point G of the rod, where $AG = 1.5$ m. The rod hangs horizontally, in equilibrium, from strings attached at A and B. The strings at A and B make angles with the vertical of $\alpha°$ and $15°$ respectively. The tension in the string at B is TN (see diagram). Find

i the value of T,

ii the value of α.

(Cambridge International AS and A level Mathematics 9709/05 Paper 5
Q5 November 2006)

39

Chapter 3
Centre of mass

This chapter shows how to find the centre of mass for objects made up of several simple parts. When you have completed it, you should

■ be able to find the centre of mass for objects made up of parts whose centres of mass you already know

■ understand how the procedure is justified by the theory of moments

■ know the weighted mean formula for finding centres of mass

■ know how to use the centre of mass to determine equilibrium positions for objects standing on a surface, hanging from a hook or suspended by a string.

3.1 One-dimensional objects

Recall the discussion of centres of mass in section 2.2 (page 21). In this chapter, you will learn how to calculate the centre of mass of some objects that are not uniform. The ideas involved are illustrated by the following example.

EXAMPLE 3.1.1

A portable radio has a telescopic aerial, consisting of three parts of masses 0.05 kg, 0.03 kg and 0.02 kg, each of length 20 cm. The first part is hinged to the radio at one end, and the other two parts slide inside it. All three parts are uniform. Find the distance of the centre of mass from the hinge when

a the aerial is closed up,

b the second and third parts are pulled out together to make an aerial of length 40 cm,

c the aerial is fully extended.

Suppose that the aerial is pointed in a horizontal direction. Then the weight of the aerial has a moment about the hinge. You can calculate this moment either by taking the parts separately, or by supposing that the total weight acts along a line through the centre of mass.

a When it is closed up, as in Fig. 3.1a, the aerial is in effect a single object of mass 0.1 kg with its centre of mass at its geometrical centre, 10 cm from the hinge.

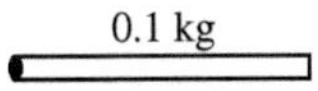

0.1 kg

Fig. 3.1a

b With the two inside parts pulled out together, as in Fig. 3.1b, the aerial consists of two sections, each of mass 0.05 kg and length 20 cm. Although not geometrically symmetrical, this is in effect a single uniform rod of mass 0.1 kg and length 40 cm, with its centre of mass 20 cm from the hinge.

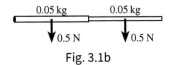

Fig. 3.1b

You can check this by considering the moments of the weights separately. Treated as two sections, with weights 0.5 N and centres of mass 10 cm and 30 cm from the hinge, the total moment is $0.5 \times 10 + 0.5 \times 30$ N cm. A single rod of weight 1 N at 20 cm from the hinge has a moment of 1×20 N cm. Either way, the calculation gives a moment of 20 N.

c Fig. 3.1c shows the aerial fully extended. You can't now appeal to symmetry to find the centre of mass, so you must use the moment argument. The three sections have weights of 0.5 N, 0.3 N and 0.2 N, acting at distances of 10 cm, 30 cm and 50 cm from the hinge. The total moment of these weights is $0.5 \times 10 + 0.3 \times 30 + 0.2 \times 50$ N cm, that is 24 N cm.

This can be equated to the moment of a weight of 1 N acting through the centre of mass. The centre of mass must therefore be $\frac{24}{1}$ cm, or 24 cm, from the hinge.

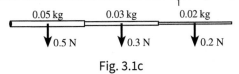

Fig. 3.1c

The calculation in part **c** of this example can be written in a more general form using algebra. Suppose that the masses of the three sections are denoted by m_1, m_2 and m_3, and that the centres of mass are at distances x_1, x_2 and x_3 from the hinge. Then the total moment of the weights about the hinge is $(m_1g)x_1 + (m_2g)x_2 + (m_3g)x_3$, which you can write more simply as $(m_1x_1 + m_2x_2 + m_3x_3)g$.

Let M denote the total mass $m_1 + m_2 + m_3$. It is usual to denote the coordinate of the centre of mass (in this case the distance from the hinge) by $\bar{x}$. The moment of the total weight can then be written as $(Mg)\bar{x}$, or $(M\bar{x})g$. Since these have to be equal,

$$\left(M\bar{x}\right)g = \left(m_1x_1 + m_2x_2 + m_3x_3\right)g$$

Cancelling g, and then dividing by M, gives

$$\bar{x} = \frac{m_1x_1 + m_2x_2 + m_3x_3}{M}$$

This expression for $\bar{x}$ is called a **weighted mean**. It is an average of the distances x_1, x_2 and x_3 in which the distances are 'weighted' according to the masses at the corresponding points.

You might already have used weighted means in statistics. For example, if in a charity collection f_1 people give $\$x_1$, f_2 people give $\$x_2$ and f_3 people give $\$x_3$, the average donation is
$$\$\frac{f_1x_1 + f_2x_2 + f_3x_3}{N},$$ *where N is the total number of contributors, so that $N = f_1 + f_2 + f_3$. This is the same formula, but in a quite different context.*

> If an object is made up of n sections of masses m_1, m_2, ... , m_n, each with its centre of mass on a line and having coordinates x_1, x_2,..., x_n, then the centre of mass has coordinate $\bar{x}$ where
>
> $$\bar{x} = \frac{m_1x_1 + m_2x_2 + ... + m_nx_n}{M} \text{ and } M = m_1 + m_2 + ... + m_n$$

It is often convenient to set out centre of mass calculations in tabular form, as in Table 3.2. On the left of the vertical line are the separate masses and distances, and on the right is the total mass at the centre of mass.

Table 3.2

Mass	m_1	m_2	m_3	...	m_n	M
Distance	x_1	x_2	x_3	...	x_n	$\bar{x}$

For example, Table 3.3 gives a summary of the data for Example 3.1.1(c).

Table 3.3

Mass (kg)	0.05	0.03	0.02	0.1
Distance (cm)	10	30	50	$\bar{x}$

From this you can write down at once the equation
$$0.05 \times 10 + 0.03 \times 30 + 0.02 \times 50 = 0.1\bar{x},$$

so $\bar{x} = \dfrac{2.4}{0.1} = 24$.

Before leaving Example 3.1.1, there is one further point to notice. In the solution the aerial was placed horizontally, but this wasn't necessary. If it had been at an angle α to the horizontal, then the moment of the weight $m_1 g$ would have been calculated (using either Fig. 3.4 or 3.5, whichever you find easier) as $(m_1 g) x_1 \cos x_1 \alpha$ or as $(m_1 g \cos \alpha) x_1$, and similarly for the weights $m_2 g$, $m_3 g$ and Mg.

Fig. 3.4   Fig. 3.5

In simplifying the equation, you would then have cancelled out the factor $g \cos \alpha$ rather than just g, and the final result would have been the same as before. It is an important property of a rigid object that, if it is moved around in space, the position of the centre of mass relative to the object remains the same.

3.2 Two-dimensional objects

If you have an object which extends in two dimensions, you will need two coordinates to describe the position of the centre of mass, and of the parts which make it up. For a rigid object you can choose a pair of axes which are fixed relative to the object, so that, as the object moves, the axes move with it.

The formulae giving $(\bar{x}, \bar{y})$, the coordinates of the centre of mass, are then a direct extension of the formula for one-dimensional objects.

> If an object is made up of n sections of masses m_1, m_2, ... , m_n, each with its centre of mass in a plane and having coordinates (x_1, y_1), (x_2, y_2), ..., (x_n, y_n), then the centre of mass has coordinates $(\bar{x}, \bar{y})$, where
>
> $$\bar{x} = \frac{m_1 x_1 + m_2 x_2 + \ldots m_n x_n}{M}, \qquad \bar{y} = \frac{m_1 y_1 + m_2 y_2 + \ldots m_n y_n}{M}$$
>
> and $M = m_1 + m_2 + \ldots + m_n$.

It is worth noting that this can also be written using vectors.

> If the centres of mass of the sections have position vectors $\mathbf{r}_1$, $\mathbf{r}_2$, ..., $\mathbf{r}_n$, then the centre of mass has position vector $\bar{\mathbf{r}}$, where
>
> $$\bar{\mathbf{r}} = \frac{m_1 \mathbf{r}_1 + m_2 \mathbf{r}_2 + \ldots + m_n \mathbf{r}_n}{M}$$

If you write each position vector in column form, $\mathbf{r}_1 = \begin{pmatrix} x_1 \\ y_1 \end{pmatrix}$ ... and $\bar{\mathbf{r}} = \begin{pmatrix} \bar{x} \\ \bar{y} \end{pmatrix}$, and read along each line in turn, you get back to the cartesian forms in the previous box.

43

EXAMPLE 3.2.1

A letter F is drawn on graph paper, as in Fig. 3.6, and a piece of thin card is then cut out to this shape. Find the centre of mass of the card.

The letter F has no symmetry, but there are several ways of splitting it up into rectangles. One way is to split it into three rectangles, as shown by the dotted lines on Fig. 3.6.

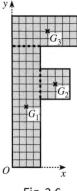

Fig. 3.6

The centres of mass of the rectangles are at the points labelled G_1, G_2 and G_3. You do not know the masses of the rectangles, but if the card is of uniform thickness the masses will be proportional to their areas. Suppose that the card has a mass of k for each small square.

Then, taking the origin at the bottom left corner of the letter, and axes across and up the page, the data can be laid out as in Table 3.7.

Table 3.7

Mass	$64k$	$16k$	$40k$	$120k$
x-coordinate	2	6	5	$\bar{x}$
y-coordinate	8	11	18	$\bar{y}$

The formulae then give

$$\bar{x} = \frac{64k \times 2 + 16k \times 6 + 40k \times 5}{120k} = \frac{53}{15},$$

$$\bar{y} = \frac{64k \times 8 + 16k \times 11 + 40k \times 18}{120k} = \frac{176}{15}.$$

The coordinates of the centre of mass are $\left(3\frac{8}{15}, 11\frac{11}{15}\right)$.

In this example, the card could be considered to be a **lamina**. This is an object which can be modelled as a plane region with no thickness.

EXAMPLE 3.2.2

A thin wire of uniform thickness is bent into a triangle with sides of length 30 cm, 40 cm and 50 cm. Find the position of the centre of mass.

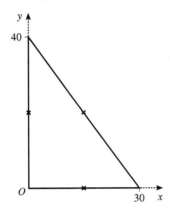

Fig. 3.8

The triangle has sides in the ratio $3:4:5$, so it is right-angled. In Fig. 3.8 the origin is taken at the right angle, with x- and y-axes along the 30 cm and 40 cm sides respectively.

Each straight section of wire has mass proportional to its length, with centre of mass at its mid-point. Let the wire have mass k kg per centimetre. Then the data are shown in Table 3.9.

Table 3.9

Mass (kg)	$30k$	$40k$	$50k$	$120k$
x-coordinate (cm)	15	0	15	$\bar{x}$
y-coordinate (cm)	0	20	20	$\bar{y}$

The formulae give

$$\bar{x} = \frac{30k \times 15 + 40k \times 0 + 50k \times 15}{120k} = 10 \, ,$$

$$\bar{y} = \frac{30k \times 0 + 40k \times 20 + 50k \times 20}{120k} = 15 \, .$$

The centre of mass is 15 cm from the 30 cm side and 10 cm from the 40 cm side.

Exercise 3A

1 A stepped rod has three uniform sections, each of length 40 cm. The masses of the sections are 0.6 kg, 0.3 kg, and 0.1 kg, as shown in the diagram. Find the distance of the centre of mass from the heavier end of the rod.

0.6 kg	0.3 kg	0.1 kg
40 cm	40 cm	40 cm

2 The frame shown in the diagram is made by welding a uniform rod, of mass 1 kg and length 1.2 m, to a uniform circular hoop, of mass 3.6 kg and radius 0.6 m. Find the distance of the centre of mass of the frame from the rod.

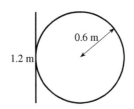

3 For some astronomical purposes it is useful to consider the earth and the moon together as a single system. Given that the mass of the earth is 82.45 times the mass of the moon, and that the mean distance of the moon from the earth is 384 400 km, find the distance of the centre of mass of the earth–moon system from the centre of the earth.

4 A piece of card is cut to form the shape shown in the diagram. Find the distance of its centre of mass from *AB*.

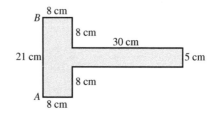

5 A badminton racket consists of a frame 28 cm long attached to a shaft of length 38 cm. For the last 16 cm of its length the shaft is surrounded by the grip. If the frame, the shaft and the grip have mass 70 grams, 40 grams and 30 grams respectively, and the centre of mass of each part is at its geometrical centre, find the distance of the centre of mass of the racket from the end of the shaft.

6 A fishing rod is made of three parts clipped together, each uniform and of length 1.2 metres. The separate parts have mass 60 grams, 40 grams and 25 grams. Half of the heaviest part is wrapped in a sleeve to which a reel is attached, with total mass 75 grams and centre of mass 40 cm from the end of the rod. Find the distance of the centre of mass of the whole rod from this end.

7 The container shown in the figure is made of sheet metal of uniform thickness. It is open at the top. Find the height of the centre of mass above the base.

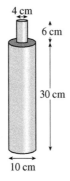

8   An oar consists of a uniform shaft of length 3 metres and a blade of length 70 cm, whose centre of mass is 40 cm beyond the shaft. The oar passes through a metal ring 80 cm from the end of the shaft. If the mass of the shaft is 4 kg, and the mass of the blade is 0.3 kg, find the distance of the centre of mass of the oar from the ring.

9   The diagram shows a shape cut from a uniform sheet of thin metal. Determine the distance of the centre of mass of this shape from the side AB and from the side AC of the shape.

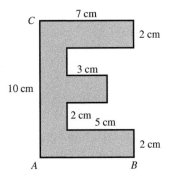

10 A frame, in the form of the quadrilateral shown in the diagram, is made from a piece of uniform wire of length 54 cm. Find the distance of the centre of mass of the frame from

    a   the side of length 18 cm,
    b   the side of length 12 cm.

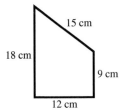

11 A pair of trousers on a hanger is modelled as a rectangular sheet of cloth attached to a horizontal rod. Another two rods, which are connected to each other, are attached to the ends of the horizontal rod, as shown in the diagram. All three rods and the sheet of cloth are uniform. The mass of the cloth is 550 grams and its centre of mass is 30 cm below the horizontal rod. The mass of the horizontal rod is 100 grams and its length is 40 cm. Each of the other rods has mass 75 grams and length 25 cm. Find the distance of the centre of mass of the combined hanger and trousers, below the horizontal bar of the hanger.

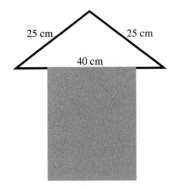

12 A stone sculpture consists of a $2\,\text{m} \times 1.6\,\text{m} \times 1\,\text{m}$ cuboid surmounted by a cylinder of radius x metres and length $1\,\text{m}$, as shown in the diagram. The centre of mass of the sculpture is a point of contact between the two parts. Find x.

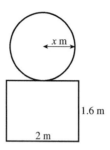

13 One end of a circular cylinder of mass $0.8\,\text{kg}$, radius $3\,\text{cm}$ and length $25\,\text{cm}$ is attached to a circular disc of mass $0.45\,\text{kg}$ and radius $15\,\text{cm}$. The axis of the cylinder is at right angles to the disc and $10\,\text{cm}$ from its centre, as shown in the diagram. Find the distance of the centre of mass of the combined cylinder and disc from

a the plane of the disc,

b the axis of the disc.

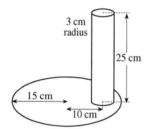

14 An ice-hockey stick is modelled as a pair of parallelograms, with dimensions as shown in the figure. Taking the mass per unit area to be constant, find the distances of the centre of mass of the stick above and to the right of the heel H.

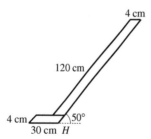

15 C is the centre of mass of the uniform piece of plywood shown in the diagram. Given that $\bar{y} = 3$, find z and $\bar{x}$.

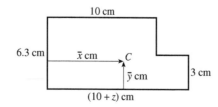

16 A sheet of metal has uniform thickness and uniform density. The shape shown in the diagram is cut from the sheet of metal. Find the coordinates $(\bar{x}, \bar{y})$ of the centre of mass C of the shape.

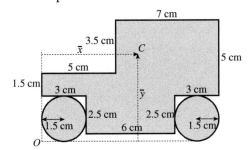

3.3 Hanging and balancing

Take the aerial in Example 3.1.1 and extend it. Suppose that it could be unhooked from the hinge and balanced on your finger. Where along the aerial would it balance?

The answer is, of course, at the centre of mass. Fig. 3.10 shows the situation with the finger 24 cm from the thick end. The weight of the 0.05 kg section has an anticlockwise moment of $0.5 \times 14 \, \text{N cm}$; the weights of the other two sections have a total clockwise moment of $(0.3 \times 6 + 0.2 \times 26) \, \text{N cm}$. So there are moments of 7 N cm both anticlockwise and clockwise, and the aerial is in equilibrium.

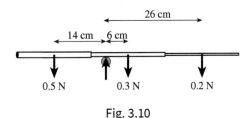

Fig. 3.10

If you have any object standing on a surface in equilibrium, there are only two forces acting: the weight acting vertically through the centre of mass, and the resultant contact force from the surface. These must therefore act along the same line, so that the centre of mass must be directly above a point at which the resultant contact force can act.

For example, if the contact is at a single point, the centre of mass must be above that point. If the contact is bounded by a circle, the centre of mass must be above a point inside or on the boundary of the circle; similarly for a rectangle or a triangle.

EXAMPLE 3.3.1

Fig. 3.11 shows a design for a piece of table sculpture which is to be carved out of a single uniform piece of marble. The dimensions are in centimetres. What is the largest possible value for the length labelled d?

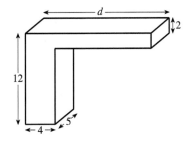

Fig. 3.11

Although the sculpture is three dimensional, it has a vertical plane of symmetry in which the centre of mass must lie. Fig. 3.12 shows the cross section in this plane of symmetry. The dotted line suggests one way of splitting the sculpture into two cuboids. If the marble has density k kg per cm^3, the mass of the upright column is $(4 \times 10 \times 5)k$ kg, and the mass of the flat slab is $(2 \times d \times 5)k$ kg. The data are summarised in Table 3.13.

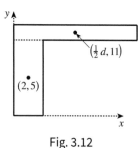

Fig. 3.12

Table 3.13

Mass (kg)	$200\,k$	$10\,dk$	$(200+10d)k$
x-coordinate (cm)	2	$\frac{1}{2}d$	$\bar{x}$
y-coordinate (cm)	5	11	$\bar{y}$

The sculpture will stand unsupported if the centre of mass lies above a point of the rectangular base, that is if $\bar{x} \leq 4$. The value of $\bar{y}$ doesn't matter. The formula gives

$$\bar{x} = \frac{400k + 5d^2k}{(200+10d)k} = \frac{80 + d^2}{40 + 2d}$$

The value of d must therefore satisfy the condition

$$\frac{80 + d^2}{40 + 2d} \leq 4, \quad \text{that is} \quad d^2 - 8d \leq 80.$$

To complete the square, add 16 to both sides, which gives $(d-4)^2 \leq 96$, so that $d \leq 4 + \sqrt{96} = 13.8$, correct to 3 significant figures.

The maximum length of the sculpture is about 13.8 cm.

EXAMPLE 3.3.2

A plastic container has the shape of a cylindrical tube, open at one end, whose height is twice the diameter of its base. The sides and the base have the same thickness. It is placed on a rough surface at 20° to the horizontal. Will it topple over if placed

a the right way up, **b** upside down?

The container is made of thin plastic, so it can be modelled as a surface of negligible thickness. If the base has radius r, the height of the container is $2(2r) = 4r$, so the area of the curved surface is $(2\pi r) \times 4r = 8\pi r^2$, which is 8 times the area of the base. So if the base has mass m, the curved surface has mass $8\,m$.

The centre of mass lies on the axis of symmetry, so you only need to find the height above the base. The data are summarised in Table 3.14.

Table 3.14

Mass	m	$8m$	$9m$
Height above base	0	$2r$	$\bar{y}$

So $\bar{y} = \dfrac{16mr}{9m} = \dfrac{16}{9}r$

a Fig. 3.15 shows the container on the sloping surface the right way up. The vertical line through the centre of mass meets the base of the container at a distance $(\frac{16}{9}r)\tan 20°$ from the centre, which is about $0.65r$. Since this is less than r, the contact force from the surface can act along this line, so that the container will not topple over. (Note that the contact force has been omitted from Fig. 3.15.)

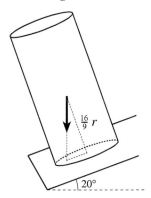

Fig. 3.15

b With the container upside down, the centre of mass is $\frac{20}{9}r$ above the rim. When it is placed on the sloping surface, the vertical line through the centre of mass meets the plane of the rim at a distance $(\frac{20}{9}r)\tan 20°$ from the centre, which is about $0.81r$. The container still will not topple over; the resultant contact force can act along this line, even though it doesn't pass through a point where the container is physically in contact with the surface. It is enough for the line of action to meet the surface at a point inside the rim of the container.

In this example the contact must also be rough enough for the container not to slide down the surface. You already know that this means that the coefficient of friction has to be more than $\tan 20°$. See M1 Experiment 5.3.2.

The same principle applies when an object hangs from a hook. The weight is then opposed by the contact force from the hook. So in equilibrium the object rests with its centre of mass directly below the hook.

EXAMPLE 3.3.3

The triangular wire in Example 3.2.2 hangs from a hook at the sharpest corner. Find the angle which the 40 cm side makes with the vertical.

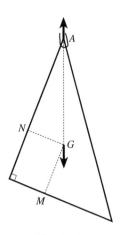

In Fig. 3.16 G is the centre of mass. From Example 3.2.2 you know that $GN = 10$ cm and $GM = 15$ cm, so that $AN = 25$ cm. Since G must be directly below A, the angle required is GAN, which is equal to $\tan^{-1}\dfrac{10}{25} = 21.8...°$.

The 40 cm side hangs at about 22° to the vertical.

Fig. 3.16

A similar situation is when the object hangs by a string. The upward force is then provided by the tension in the string, so the centre of mass must lie on the line of the string produced.

Sometimes it is quite difficult to draw diagrams with the hanging object at an angle. The next example illustrates a useful trick which can be used to get round this difficulty.

EXAMPLE 3.3.4

An open box has five square faces of side 10 cm, each having mass m kg. Initially it stands with its base on the floor. A string is then fixed to one of the upper corners, and the box is lifted by the string until it is clear of the floor. Find the angle now made with the vertical by the edges which were originally vertical.

Fig. 3.17 shows the box in its initial position. The four vertical faces have a total mass of $4m$ kg, and their combined centre of mass is 5 cm above the centre of the base. The base has mass m kg. So the height of G, the centre of mass of the box, above the base is $\dfrac{4m \times 5 + m \times 0}{5m}$ cm, which is 4 cm.

Let C be the middle of the top of the box, and A the corner at which the string is fixed. When the box hangs from the string, the line AG is vertical. Instead of drawing the box again in its new position, you can simply join AG and label it 'vertical'. It is only a convention that vertical lines are drawn up the page! The angle wanted is equal to the angle AGC. It is easy to calculate that $GC = 6$ cm and $AC = 5\sqrt{2}$ cm,

so this angle is $\tan^{-1}\dfrac{5}{6}\sqrt{2} = 49.68...°$

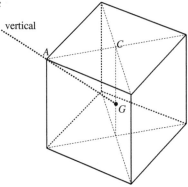

When the box hangs from A by a string, the edges are at about 50° to the vertical.

Fig. 3.17

3.4 The theory justified

It was stated at the end of Section 3.1 that, if a rigid object is moved around in space, the position of the centre of mass relative to the object remains the same. It is now time to explore this statement more precisely.

The object is rigid, so it keeps the same shape as it is moved around. This means that you can set up axes fixed in the object, which move around with it. The object is made up of parts with masses denoted by m_i (where the suffix i takes values 1, 2,..., n) located at points P_i whose coordinates remain constant as it moves.

To avoid too much complication, the proof in this section will be restricted to two dimensional objects in a vertical plane, so that you only need two coordinates (x_i, y_i).

What can then be shown is:

> In any equation of resolving or moments, the weights of the separate parts can be replaced by the weight of a mass M located at the point G with coordinates $(\bar{x}, \bar{y})$, where M, $\bar{x}$, $\bar{y}$ are defined as in Section 3.2.

The proof for resolving equations is trivial. If you are resolving in a direction making an angle θ with the vertical, the resolved parts of the weights contribute an amount $m_1 g\cos\theta + m_2 g\cos\theta + \ldots + m_n g\cos\theta$, which you can write as $(m_1 + m_2 + \ldots + m_n)g\cos\theta$, which is $Mg\cos\theta$.

To prove the result for moments equations it helps to use the trick introduced in Example 3.3.4, setting the axes in their conventional position and tilting the line representing vertical. In Fig. 3.18 the weights are drawn at an angle α to the x-axis; this corresponds to a position in which the rigid object is turned so that the x-axis makes an angle α with the vertical.

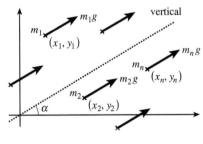

Fig. 3.18

Suppose that you are taking moments about a point H with coordinates (a, b). The moment of the weight $m_i g$ is $m_i g p_i$, where p_i denotes the perpendicular distance from H to the line of action of the weight. You can see from Fig. 3.19 that this distance is equal to $(x_i - a)\sin\alpha - (y_i - b)\cos\alpha$. So the total moment of all the weights is the sum of terms of the form

$$m_i g\big((x_i - a)\sin\alpha - (y_i - b)\cos\alpha\big)$$

which you can write as

$$m_i x_i g\sin\alpha - m_i a g\sin\alpha - m_i y_i g\cos\alpha + m_i b g\cos\alpha$$

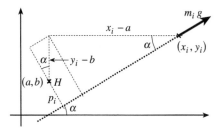

Fig. 3.19

Adding up these moments for 1, 2, ..., n and replacing $m_1 + m_2 + ... + m_n$ by M, $m_1 x_1 + m_2 x_2 + ... + m_n x_n$ by $M\bar{x}$, and $m_1 y_1 + m_2 y_2 + ... + m_n y_n$ by $M\bar{y}$, this becomes

$$M\bar{x}g\sin\alpha - Mag\sin\alpha - M\bar{y}g\cos\alpha - Mbg\cos\alpha,$$

which is $Mg\left((\bar{x} - a)\sin\alpha - (\bar{y} - b)\cos\alpha\right)$, the moment of the force Mg at an angle α acting through G.

This completes the proof. The property is so basic in mechanics that **G** was once commonly called the **centre of gravity**. But since the point **G** has other properties which continue to hold even where there is no gravity, it is better to stick with the term centre of mass.

Exercise 3B

1 Find by experiment the position of the centre of mass of

   a a cricket bat,   **b** a squash racket,   **c** a snooker cue.

2 A straight rigid rod AB is suspended by a string attached at a point P of its length. The centre of mass of the rod is at G. Describe how the rod would hang

   a if P is between A and G,

   b if P is between G and B,

   c if P is at G.

3 The stepped rod of Question 1 of Exercise 3A is held in a horizontal position, suspended by two vertical strings. One of the strings is attached to the end A of the rod, as shown in the diagram. Find the tension in each of the strings when the other string is attached to the rod

   a at its centre of mass,   **b** at its end B.

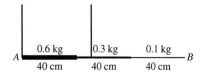

4 Carry out the following experiment. Begin by standing to attention. Then try to carry out the following movements, keeping the rest of your body upright.

   a Raise your right arm sideways to a horizontal position.

   b Raise your right leg sideways as far as you can.

   c Raise your right arm and your right leg sideways at the same time.

   d Raise your left arm and your right leg sideways at the same time.

Describe and explain what happens in each case.

5 A thick sheet of plywood in the shape of a parallelogram has sides of length 30 cm and 50 cm, with an angle of $\alpha°$ between them. What can you say about α if the sheet can stand upright in equilibrium with one of its shorter edges on a horizontal surface?

6 The structures in Question 10 and Question 2 of Exercise 3A are hung in equilibrium from a string as shown in the following diagrams. Find the angles α, β, γ and δ.

a

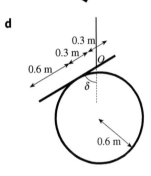

b

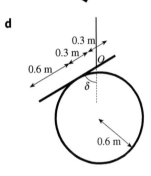

c

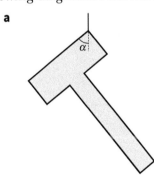

d

7 The lamina in Question 4 of Exercise 3A is suspended in equilibrium as shown in the following diagrams. Find the angles α, β, γ and δ.

a

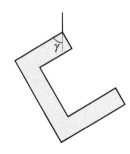

b

c

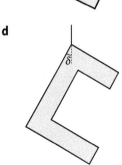

d

8 The combined cylinder and disc in Question 13 of Exercise 3A is suspended as shown in the accompanying diagram. Find the angle α.

9 A walking stick is made in three straight sections of the same diameter, with dimensions as shown in the figure. What angle will the upright part of the stick make with the vertical if it is supported in the palm of a person's hand

a at A,   **b** at B?

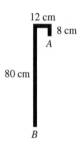

12 cm
8 cm
A
80 cm
B

56

10 An L-shaped block of uniform thickness has the corners of one of its faces labelled A, B, C, D, E and F, as shown in the diagram. The block stands on a plane which is inclined at an angle α to the horizontal. In which of the following configurations P, Q, R and S is the block on the point of toppling when $\tan \alpha$ is

a $\frac{1}{7}$,   **b** $\frac{5}{7}$,   **c** 1,   **d** $\frac{7}{5}$?

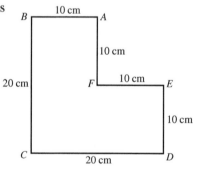

P AB coincides with a line of greatest slope of the plane with B above A.

Q BC coincides with a line of greatest slope of the plane with C above B.

R CD coincides with a line of greatest slope of the plane with D above C.

S DE coincides with a line of greatest slope of the plane with E above D.

11 An L-shaped block of uniform thickness 10 cm has dimensions as shown in the first diagram. Two cylindrical 'handles', each of radius 4 cm and length 40 cm and made of the same material as the block, are attached so that they have a common axis which is at right angles to the face $ABCDEF$, as shown in the second diagram. This axis is 5 cm from AB and 5 cm from AF. Determine whether the block topples when placed on a horizontal surface with BC vertical, and DC in contact with the surface.

Find whether the result would be the same if the length of the cylinders was 39.5 cm.

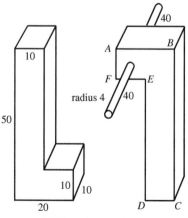

All dimensions are in cm

Miscellaneous exercise 3

1 The structure shown in the diagram, consisting of a right-angled triangle inscribed in a circle, is made of uniform wire. Calculate the distance of the centre of mass of the structure from

a *AB*, **b** *BC*.

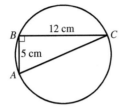

2 A uniform lamina is made in the shape shown.

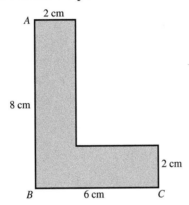

Calculate the distance of the centre of mass of this lamina from

a *AB*, **b** *BC*.

3 A walking stick consists of a pole of length 1 m and mass 1.5 kg. At the top end of the stick, a decorative ball is fastened, which has diameter 8 cm and mass 200 g. Find the distance of the centre of mass of the walking stick from the bottom end.

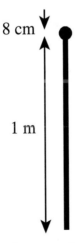

4 A signpost is made up of a wooden frame and a sign. The wooden frame consists of a thin vertical post 2.5 m high with a thin horizontal beam of length 0.8 m attached to the top of the post. The post and beam are both uniform, and their masses are proportional to their lengths. The entire wooden frame has mass 12.5 kg. Attached to the frame is a uniform rectangular sign with length 0.8 m, height 0.5 m and mass 1.5 kg, as shown in the diagram.

57

Find the distance of the centre of mass of the signpost from the vertical post and from the horizontal beam.

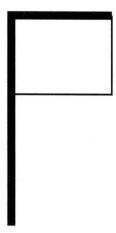

5 Two uniform cylinders, made from the same material, are arranged one on top of the other as shown in the diagram. The radius of the upper cylinder is half that of the lower, and the height of the upper cylinder is twice that of the lower. The overall height of the structure is H. Find, in terms of H, the distance of the centre of mass from the base.

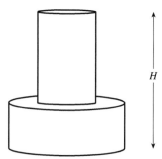

6 A structure made of a uniform wire consists of a circular piece, of radius r, and a straight piece. The straight piece forms a chord AB of the circle, which subtends an angle of 120° at the centre O of the circle, as shown in the diagram. Show that the distance, from the chord, of the centre of mass of the structure is $\dfrac{\pi r}{2\pi + \sqrt{3}}$.

The structure is suspended from the point A. Find the angle that the chord makes with the downward vertical.

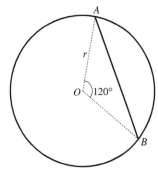

7 Two blocks, one rectangular and one L-shaped, are of uniform thickness and made of the same material. The dimensions of the blocks are as shown in the diagram. Find the value of x when

a the blocks are unattached and the L-shaped block is on the point of toppling from the rectangular block,

b the blocks are attached and the combined blocks are on the point of toppling.

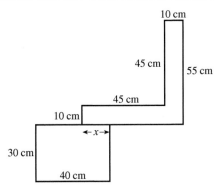

8 A uniform wire is bent into a right-angled triangle *ABC* with sides *AB* = 5 cm, *BC* = 12 cm and *AC* = 13 cm.

The wire hangs in equilibrium from a string attached to *A*. Find the angle between *AB* and the vertical.

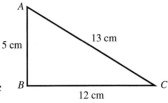

9 A uniform solid cuboid is placed at rest on a rough plane. The dimensions of the cuboid are 10 cm by 10 cm by *h* cm, and the orientation of the cuboid is as shown in the diagram. The plane is inclined at 20° to the horizontal, and the sides of length *h* cm are parallel to a line of greatest slope on the plane. The coefficient of friction between the cuboid and the plane is *μ*.

Given that the cuboid remains in equilibrium, find the minimum possible values of *h* and of *μ*.

10 A uniform square lamina *ABCD* has side length *l* and mass *M*. A particle of mass *m* is attached to the lamina at *B*. The lamina is freely suspended from point *A* and hangs in equilibrium. The side *AB* makes an angle of *θ* with the downward vertical. Find an expression for $\tan \theta$ in terms of *l*, *M* and *m*.

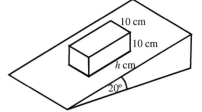

11 A mug is placed on a rough surface inclined at *α* to the horizontal. *ABCH* is modelled as a uniform hollow cylinder of mass 200 g with a uniform circular base of mass 50 g. The handle *DEFG* is modelled as a uniform rectangular frame. The four sides have total mass 50 g and the handle lies in the direction of greatest slope of the plane, as shown in the diagram.

The dimensions of the mug are *AB* = 9.5 cm, *AH* = 8 cm, *CD* = 2 cm, *DG* = 6 cm and *DE* = 3 cm.

a Find the distance of the centre of mass of the mug from

   i *AH*,   **ii** *CH*.

b Given that the cup remains in equilibrium and that the surface is rough enough to prevent sliding, determine the maximum possible value of *α*.

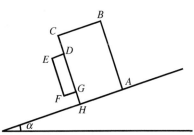

12 Toy bricks, all uniform cubes of the same size, are contained in a box. With the box open a child builds a column of the bricks in its lid, as shown in the diagram. With eight bricks used the column does not topple, but when the ninth brick is added the column topples. Show that the angle the lid makes

with the horizontal lies between $\tan^{-1}\dfrac{1}{9}$ and $\tan^{-1}\dfrac{1}{8}$.

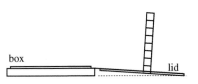

13

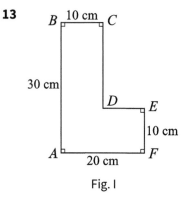

Fig. I

ABCDEF is the L-shaped cross-section of a uniform solid. This cross-section passes through the centre of mass of the solid and has dimensions as shown in Fig. I.

i Find the distance of the centre of mass of the solid from the edge *AB* of the cross-section.

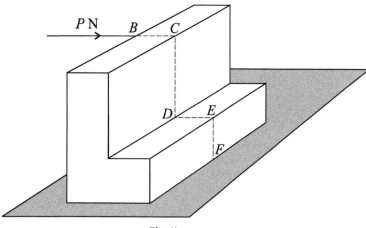

Fig. II

The solid rests in equilibrium with the face containing the edge *AF* of the cross-section in contact with a horizontal table. The weight of the solid is W N. A horizontal force of magnitude P N is applied to the solid at the point *B*, in the direction of *BC* (see Fig. II). The table is sufficiently rough to prevent sliding.

ii Find P in terms of W, given that the equilibrium of the solid is about to be broken.

(Cambridge International AS and A level Mathematics 9709/05
Paper 5 Q3 June 2005)

14 A uniform lamina of weight 15 N has dimensions as shown in the diagram.

i Show that the distance of the centre of mass of the lamina from *AB* is 0.22 m. [4]

The lamina is freely hinged at *B* to a fixed point. One end of a light inextensible string is attached to the lamina at *C*. The string passes over a fixed smooth pulley and a particle of mass 1.1 kg is attached to the other end of the string. The lamina is in equilibrium with *BC* horizontal. The string is taut and makes an angle of $\theta°$ with the horizontal at *C*, and the particle hangs freely below the pulley (see diagram).

ii Find the value of θ.

(Cambridge International AS and A level Mathematics
9709/05 Paper 5 Q5 June 2006)

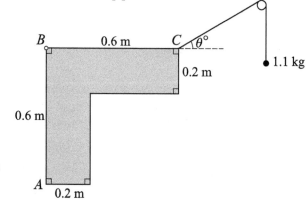

Chapter 4
Rigid objects in equilibrium

This chapter establishes the conditions for a rigid object to be in equilibrium under the action of a number of coplanar forces. When you have completed it, you should

- know and be able to apply general conditions for the equilibrium of a rigid object
- be able to determine how equilibrium is broken as a force is increased
- be able to find lines of action of forces needed to maintain equilibrium
- recognise that some problems are indeterminate and do not have a unique solution.

4.1 Equilibrium equations

Suppose that a ladder is put up against the wall of a building, and that both the ground and the wall are so smooth that friction can be neglected. You would expect the ladder to slide down until it lies flat on the ground.

Fig. 4.1 explains in terms of the forces on the ladder why it can't rest against the wall in equilibrium. There are only three forces: the weight, and the normal contact forces (R and S) from the ground and the wall. Although the forces would balance vertically if R were equal to W, there is no horizontal force to counteract S.

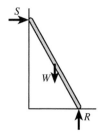

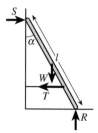

Fig. 4.1   Fig. 4.2

For the ladder to be in equilibrium there has to be some other force with a horizontal resolved part.

This could be provided for example by attaching a rope from the ladder to an anchor on the wall. Fig. 4.2 shows the forces with the addition of a tension in a horizontal rope attached to the ladder one-quarter way up its length.

You could now write down two equations, resolving horizontally and vertically:

$$\mathcal{R}(\rightarrow)\ S = T, \quad \text{and} \quad \mathcal{R}(\uparrow)\ R = W.$$

The weight of the ladder is presumably known, so you know the value of R. But you don't know either S or T, so you need a third equation to find these. This is provided by an equation of moments about some point. You can choose any point you like, for example the foot of the ladder.

Denote the length of the ladder by l and the angle with the vertical by α.

$$\mathcal{M}(\text{foot of ladder})\quad S \times l\cos\alpha = W \times \tfrac{1}{2}l\sin\alpha + T \times \tfrac{1}{4}l\cos\alpha.$$

Dividing through by l and substituting $S = T$ from the $\mathcal{R}(\rightarrow)$ equation,

$$T\cos\alpha = \tfrac{1}{2}W\sin\alpha + \tfrac{1}{4}T\cos\alpha,$$

so

$$\tfrac{3}{4}T\cos\alpha = \tfrac{1}{2}W\sin\alpha$$

It follows that $T = \tfrac{2}{3}W\tan\alpha$. Since $S = T$, all three unknown forces have now been found in terms of W and α.

This is a typical way of finding the forces which keep a rigid object in equilibrium. You can write down two resolving equations and one moments equation, and then solve these for the unknown forces.

EXAMPLE 4.1.1

A rectangular stone slab $ABCD$ has edges $AB = 5\,\text{m}$ $BC = 3\,\text{m}$. It is lying horizontally, and being manoeuvred into position by horizontal forces at A, B, C and D directed along the edges, as shown in Fig. 4.3. The force at A has magnitude 75 N, and the forces at B, C and D are X, Y and Z newtons respectively. Calculate the values of X, Y and Z needed to keep the slab in equilibrium.

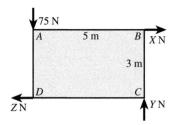

Fig. 4.3

Two resolving equations are

$$\mathcal{R}(\rightarrow)X - Z = 0, \qquad \mathcal{R}(\uparrow)Y - 75 = 0$$

These give the value of Y, and show that X and Z are equal. But to find the actual values of X and Z you need one more equation, which must be an equation of moments. There is no obvious point to take moments about, so choose B. The forces X and Y have no moment about B, so

$$\mathcal{M}(B)\quad 75 \times 5 - Z \times 3 = 0.$$

This gives $Z = 125$, and it follows that $X = 125$.

The forces at B, C and D are 125 N, 75 N and 125 N respectively.

Choose some other point (not necessarily a corner of the slab) and check for yourself that the moments of the four forces balance about that point.

EXAMPLE 4.1.2

A student has on her desk a glass paperweight in the shape of a hemisphere of radius 4 cm. She rests a uniform pen 12 cm long against it, at 30° to the horizontal, with its lower end on the desk, as shown in Fig. 4.4. The paperweight does not move. If the contact between the pen and the paperweight is smooth, how large must the coefficient of friction between the pen and the desk be to maintain equilibrium?

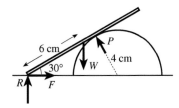

Fig. 4.4

You are not given the weight of the pen, so denote it by W. The other forces on the pen are the normal contact force P from the paperweight, and the friction F and normal contact force R from the desk.

It is simplest to begin by taking moments about the end of the pencil on the desk, which will give P in terms of W. You can then find F and R by resolving horizontally and vertically.

The distance from the end of the pen to its point of contact with the paperweight is $\dfrac{4}{\tan 30°}$ cm, which is $4\sqrt{3}$ cm; the distance to the line of action of the weight, measured horizontally, is $6\cos 30°$ cm, which is $3\sqrt{3}$ cm. So, using centimetre units for length,

$$\mathcal{M}(\text{end of pencil}) \quad P \times 4\sqrt{3} = W \times 3\sqrt{3}, \quad \text{giving } P = \tfrac{3}{4}W$$

$$\mathcal{R}(\rightarrow)\ F = P\cos 60°, \quad \text{so } F = \tfrac{1}{2}P = \tfrac{3}{8}W$$

$$\mathcal{R}(\uparrow)\ R + P\cos 30° = W, \quad \text{so } R = W - \tfrac{3}{4}W \times \tfrac{1}{2}\sqrt{3} = W\left(1 - \tfrac{3}{8}\sqrt{3}\right).$$

If μ denotes the coefficient of friction between the pen and the desk, it is necessary that $F \leq \mu R$ for equilibrium to be possible, so

$$\tfrac{3}{8}W \leq \mu W\left(1 - \tfrac{3}{8}\sqrt{3}\right),$$

$$\mu \geq \frac{\tfrac{3}{8}}{1 - \tfrac{3}{8}\sqrt{3}} = \frac{3}{8 - 3\sqrt{3}} = 1.07, \text{ correct to 3 significant figures.}$$

For the pen to rest in equilibrium against the paperweight at an angle of $30°$, the coefficient of friction between the pen and the desk must be at least 1.07.

EXAMPLE 4.1.3

A high diving board is a uniform horizontal plank of length 2 metres and weight $100\,\text{N}$. It is hinged to the step tower at one end, and supported at 0.5 metres from that end by a strut which produces a thrust at $20°$ to the vertical. Find the magnitude of the thrust and the force from the hinge when a diver of weight $700\,\text{N}$ stands at the other end of the board.

It would seem that the hinge has to pull the board in towards the tower and also exert a downward force on the board. In Fig. 4.5 the force from the hinge is therefore towards the left and downwards. It is simplest to find these as two components, X newtons and Y newtons, and then to combine them. Let the thrust in the strut be S newtons.

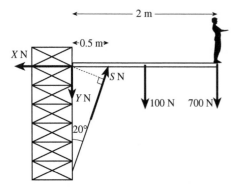

Fig. 4.5

The natural choices for the three equations are to take moments about the hinge and to resolve in horizontal and vertical directions.

$\mathcal{M}(\text{hinge})$ $\qquad S(0.5\cos 20°) = 100 \times 1 + 700 \times 2.$

$\mathcal{R}(\rightarrow)$ $\qquad\qquad S\cos 70° = X.$

$\mathcal{R}(\uparrow)$ $\qquad\qquad S\cos 20° = Y + 100 + 700.$

Working to 4 significant figures, the first equation gives $S = 3193$, and then $X = 1092$ and $Y = 2200$. Finally, X and Y can be combined, using Fig. 4.6, to show that the force from the hinge is 2456 N at an angle of 63.60° to the horizontal.

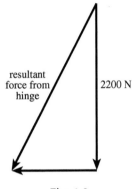

resultant force from hinge

2200 N

Fig. 4.6

So the thrust from the strut is about 3190 N, and the force from the hinge is about 2460 N at 63.6° to the horizontal. If at the start, the force Y N had been drawn pointing upwards, then the calculations would have led to a negative value for Y, showing that the force actually acts upwards.

In Example 4.1.3, if you do the calculations for yourself, you will find that the value $Y = 2200$ is exact. This is because another way of calculating Y is to take moments about the point where the strut meets the board, giving the equation $0.5Y = 0.5 \times 100 + 1.5 \times 700$. There is always the possibility of replacing one, or even both, of the resolving equations by an equation of moments.

In the next example two equations of moments are used, so that the contact forces can each be found from a single equation.

EXAMPLE 4.1.4

A quarry truck has wheels 4 metres apart. When fully loaded with stone, the centre of mass is 1.2 metres from the rails and midway between the wheels. A brake can be used to lock one of the pairs of wheels when the truck is on a slope. Is it better for the brake to be on the lower or the upper wheels? If the coefficient of friction between the wheels and the rails is 0.4, what is the steepest slope on which the truck can stand when fully loaded?

Fig. 4.7 shows the truck on a slope of angle α. You don't yet know whether the frictional force F acts on the lower or the upper wheels. It is simpler to write the equations of moments if you split the weight W into components $W\sin\alpha$ and $W\cos\alpha$ parallel and perpendicular to the slope. The contact forces at the upper and lower wheels are R and S.

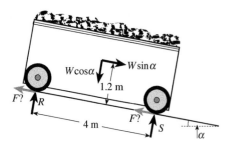

Fig. 4.7

$\mathcal{M}$(lower wheel) $4R + 1.2W\sin\alpha = 2W\cos\alpha.$

$\mathcal{M}$(upper wheel) $4S = 1.2W\sin\alpha + 2W\cos\alpha.$

So $R = W(0.5\cos\alpha - 0.3\sin\alpha)$ and $S = W(0.5\cos\alpha + 0.3\sin\alpha)$. It follows that $S > R$. Since the greatest frictional force is proportional to the contact force, you can get more friction by having the brake on the lower wheels. Let F be the frictional force.

$\mathcal{R}(\|$ to slope$)$ $F = W\sin\alpha.$

For the truck not to move down the slope, $F < 0.4S$, so that

$$W\sin\alpha < 0.2W\cos\alpha + 0.12W\sin\alpha.$$

Therefore $0.88\sin\alpha < 0.2\cos\alpha$, which gives $\tan\alpha < \dfrac{0.2}{0.88}$. The maximum angle of slope is $\tan^{-1}\dfrac{0.2}{0.88}$, which is $12.80°$, correct to 2 decimal places.

The lower wheels should be braked, and the angle of slope cannot exceed $12.8°$.

Notice in this example that, if you add the two equations of moments and divide by 4, you get $R + S = W\cos\alpha$, which is the equation of resolving at right angles to the slope.

So to get any new information by resolving, you have to choose a direction which is not at right angles to the line joining the two points about which you take moments.

For a similar reason, if you choose to take moments about a third point, you will only get new information if this point is not in line with the first two.

The equations for the equilibrium of a rigid object acted on by coplanar forces can be obtained by

- **taking moments about one point and resolving in two different directions; or**
- **taking moments about two points and resolving in a direction which is not perpendicular to the line joining them; or**
- **taking moments about three points which are not collinear.**

Exercise 4A

1 A uniform metal bar, of length 4 metres and weight 2000 newtons, is being pushed horizontally from one end across two supports 2 metres apart. The support closer to the pushing force is a light rail which can rotate smoothly about a horizontal axis. The other support is fixed, and the coefficient of friction at this support is 0.6. Calculate the force necessary to push the bar at constant speed when x metres of its length projects beyond the fixed support.

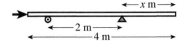

2 In Question 1 the fixed support is raised so that the metal bar is inclined at 10° to the horizontal. The distance between the two supports is still 2 metres. Find the least value of x which will enable the bar to rest in equilibrium on the two supports without any externally applied force.

3 A uniform rectangular picture $ABCD$, 3 metres wide and 2 metres high, has weight 400 newtons. It is supported on a trestle at a point E of DC, 1 metre from D, and kept upright by means of a rope attached to the picture at A. Calculate the tension in the rope and the normal and frictional forces at E

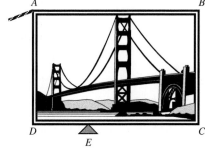

    a if the rope is horizontal,

    b if the rope is held at 40° to the vertical.

4 In a stretching exercise two men, of the same height and weight, stand toe-to-toe, each grasping the other's wrists. Both lean backwards, so that their arms are stretched, and both keep their bodies straight. One of the men is modelled as a rod having one end A in contact with the ground, which is horizontal. The weight of 800 N acts through the point 100 cm from A. The rod is kept at an angle of 66° to the horizontal by a horizontal force of magnitude P newtons applied at the point which is 150 cm from A, as shown in the diagram. Find P.

Given that the men are on the point of slipping and that their feet do not touch each other, use the model to estimate the coefficient of friction between the men's shoes and the floor.

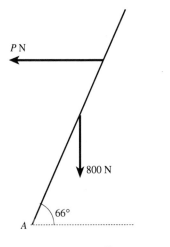

5 A rectangular platform $ABCD$ of weight 200 N is smoothly hinged, along its edge AB, to a vertical wall. The platform is kept horizontal by two parallel chains, inclined at 45° to the horizontal, connecting the points P and Q of the wall to the points D and C respectively, as shown in the diagram. P and Q are vertically above A and B respectively. A man of weight 850 N stands on the edge of the platform midway between D and C. Find

    a the tension in each of the chains,

    b the magnitude of the total force exerted on the hinge by the wall.

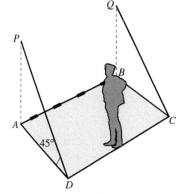

67

6 A gardener is taking a wheelbarrow down a path which slopes at 6° to the horizontal. She puts it down for a rest. The rear legs of the barrow are 120 cm behind the front wheel. The weight of the barrow and its load is 300 newtons, and the centre of mass, G, is 40 cm from the path and 50 cm behind the front wheel. There is no friction at the contact between the wheel and the path. What is the least coefficient of friction between the legs and the path which will enable the barrow to rest in equilibrium?

The gardener now lifts up the barrow and starts to wheel it down the path at a steady speed. She holds the handles at points 150 cm behind the front wheel and 60 cm from the path. Find the components parallel and perpendicular to the path of the total force from her two arms. Hence find the tension in each arm and the angle between her arm and the direction of the slope.

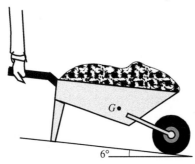

7 A uniform beam AB has length 3.5 m and weight 200 N. The beam has its end A in contact with horizontal ground and is at rest propped against a marble slab of height 0.7 m, as shown in the diagram. A is 2.4 m from the nearest face of the slab. Assuming the contact between the beam and the slab is smooth, find the normal and frictional components of the contact force at A, and hence find the least possible value of the coefficient of friction between the beam and the ground.

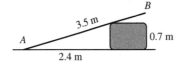

8 A car of weight W is parked facing downwards on a straight road inclined at angle α to the horizontal. The hand-brake is on, which has the effect of locking the back wheels. The total normal contact force on the front wheels is S and the total normal contact force on the back wheels is R. Given that $S + R = \dfrac{24}{25}W$, find the value of $\cos\alpha$. Find also the total frictional force, in terms of W, on the rear wheels.

The distance between the front and rear wheels is L. The distance of the centre of mass of the car behind the front wheels is kL, and its distance from the road surface is h. Show that $R = \dfrac{1}{25}W\left(24k - \dfrac{7h}{L}\right)$.

Given that $k = 0.4$ and that $\dfrac{h}{L} = 0.3$, find the ratio $\dfrac{S}{R}$.

4.2 Breaking equilibrium by sliding or toppling

Suppose that you are trying to push a box across the floor by applying a force along the top edge. If the floor is not too rough, there should be no problem so long as you push hard enough. But if the floor is very rough, the box may topple over about the opposite edge.

In practice you would probably start with quite a small force, and gradually increase it until the box starts to move. Fig. 4.8 illustrates what you hope will happen, assuming that the box has not yet toppled over. If the box has weight W and you push horizontally, the normal force is also W. You must therefore push with a force greater than the limiting friction μW.

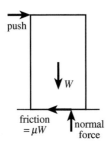

Fig. 4.8

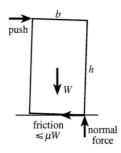

Fig. 4.9

Fig. 4.9 shows what you hope will not happen, assuming that the box has not started to slide. If the box has width b and height h, with its centre of mass on the centre vertical line, moments about the opposite edge show that the box will start to topple when you push with a force of $\dfrac{Wb}{2h}$.

What happens depends on which is smaller, μ or $\dfrac{b}{2h}$. If $\mu < \dfrac{b}{2h}$, then as you increase the force the box will start to slide; if $\mu > \dfrac{b}{2h}$, it will start to topple.

EXAMPLE 4.2.1

A small boy of weight 300 N sits on a square table, dangling his legs over the edge. His centre of mass G_B is directly above the line of the legs, 120 cm above the floor. The table has mass 20 kg, and its centre of mass G_T is 70 cm above the floor. The feet of the legs form a square of side 160 cm. The floor is very rough, and the coefficient of friction between the seat of the boy's trousers and the table is 0.3. His sister starts to tilt the table about the legs above which the boy is sitting. Will he slide off the table, or will the boy and the table topple over together?

Fig. 4.10 shows the situation before the table begins to tilt, and Fig. 4.11 shows the situation when the table is at an angle θ to the floor. Suppose that the boy is still on the table when it starts to topple over. To make the calculation of moments easier, the weights of the boy and the table have been split into components parallel and perpendicular to the table-top.

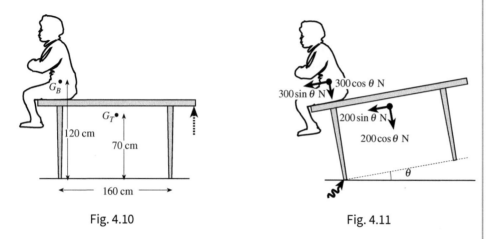

Fig. 4.10   Fig. 4.11

$\mathcal{M}(\text{feet of legs on floor})$ $300 \sin \theta \times 120 + 200 \sin \theta \times 70 = 200 \cos \theta \times 80.$

This gives $\tan \theta = \dfrac{16\ 000}{36\ 000 + 14\ 000} = 0.32.$

Now you know that, for an object on a sloping surface, sliding takes place when $\tan \theta = \mu$. (See M1 Experiment 5.3.2.) So, if the table has not already toppled, the boy will slide off the table when $\tan \theta = 0.3$.

As the table is tilted, the value of $\tan \theta$ increases from zero. It will reach the value 0.3 first, so the boy will start to slide off the table before the table topples over with him on it.

EXAMPLE 4.2.2

In a forest, after trees have been felled and trimmed, the trunks are towed to a central collection point. A cable is attached to the narrow end of each trunk, and the tension is then increased by a winch on the towing truck. One such trunk is 10 metres long, and its centre of mass is 6 metres from the end to which the cable is attached. The coefficient of friction of the trunk with the ground is 0.6. What angle should the cable make with the horizontal if the trunk is to be lifted clear of the ground rather than dragged along?

Let T_1 newtons be the tension in the cable needed to lift the end from the ground, supposing that the trunk has not started to drag. The forces are then as shown in Fig. 4.12. If the cable is at an angle α to the horizontal, and the weight of the trunk is W newtons,

$\mathcal{M}(\text{thick end})\ T_1(10 \sin \alpha) = 4W.$

That is, $T_1 = \dfrac{2W}{5 \sin \alpha}.$

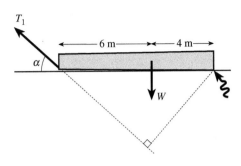

Fig. 4.12

Let T_2 newtons be the tension in the cable needed to drag the trunk along the ground, supposing that it has not started to lift clear of the ground. Friction is then limiting, and the forces are as shown in Fig. 4.13. If the normal contact force is R newtons,

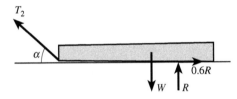

Fig. 4.13

$\mathcal{R}(\rightarrow)\ T_2\cos\alpha = 0.6R$, and

$\mathcal{R}(\uparrow)\ T_2\sin\alpha + R = W$.

Eliminating R from these equations,

$$T_2\cos\alpha = 0.6(W - T_2\sin\alpha),$$

so $T_2 = \dfrac{0.6W}{\cos\alpha + 0.6\sin\alpha}$.

As the tension in the cable is increased, the trunk will lift off the ground first if $T_1 < T_2$, that is if

$$\frac{2}{5\sin\alpha} < \frac{0.6}{\cos\alpha + 0.6\sin\alpha},$$

which gives $2\cos\alpha + 1.2\sin\alpha < 3\sin\alpha$, or $2\cos\alpha < 1.8\sin\alpha$, leading to $\tan\alpha > \dfrac{2}{1.8}$, and $\alpha > 48.0°$, correct to 3 significant figures.

So to lift the trunk clear of the ground before it starts to drag, the cable should make an angle of at least 48° with the horizontal.

4.3 Locating lines of action

In many problems you know the points at which the various forces act, but what is not known is their magnitude or their direction. However, this is not always the case.

Look back at Fig. 4.8. What is shown as the normal force from the floor on the box is in fact the resultant of small forces acting at all the points where the box and the floor are in contact. Before you start to push the box, this resultant acts at the centre of the base, directly below the centre of mass of the box. When the force is applied at the top edge, the contact forces get bigger on the right side of the base and smaller on the left side, so the resultant normal force moves to the right.

Fig. 4.14 shows the situation when you push with a horizontal force of magnitude P, before the box starts to either slide or topple. Suppose that the resultant normal force then acts at a point X on the base of the box, at a distance x from the centre. If you take moments about X, then neither the normal force nor the friction come into the equation.

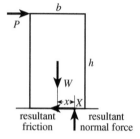

Fig. 4.14

$$\mathcal{M}(X) \quad Wx = Ph$$

It follows that $x = \dfrac{Ph}{W}$. Since h and W are constant, this shows that, as P increases, x increases and X moves to the right.

If the floor is rough enough, then x eventually reaches $\frac{1}{2}b$, when $P = \dfrac{Wb}{2h}$. The box is then just about to start to topple.

EXAMPLE 4.3.1

A raft is stuck on a hidden obstruction in a river-bed (see Fig. 4.15). Ropes are attached to one edge of the raft at A, B and C, where $AB = 6\,\text{m}$ and $CB = 2\,\text{m}$. The crew pull on these ropes with forces of 400 N, 350 N and 450 N at right angles to the edge of the raft, but fail to move it. Where is the obstruction?

Suppose that the obstruction lies along a line through X perpendicular to the edge of the raft, where X is the point of the edge x metres from A, as shown in Fig. 4.15. If you take moments about the obstruction, the force which it produces will not appear in the equation.

$\mathcal{M}(\text{obstruction})$

$$400x = 350(6 - x) + 450(8 - x).$$

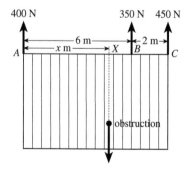

Fig. 4.15

This gives $1200x = 5700$, so $x = 4\frac{3}{4}$.

The obstruction is somewhere along the perpendicular to the edge through a point X, $4\frac{3}{4}$ metres from A.

In this example you could just as well have taken moments about any point of the edge. For example, $\mathcal{M}(A)$ would have given the equation $1200x = 6 \times 350 + 8 \times 450$

directly. To use this, you have to begin by finding the force from the obstruction by resolving, but this is not difficult!

However, an advantage of taking moments about the unknown point is that just one of the unknown quantities appears in each equation. You can see this in the next example, which has forces in two dimensions.

EXAMPLE 4.3.2

In a mechanism a rectangular plate $ABCD$ is subjected to forces at A, B and D as shown in Fig. 4.16. Where should a fourth force be applied to keep the plate in equilibrium?

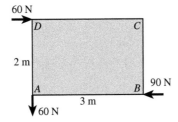

Fig. 4.16

A convenient method is to use coordinates, shown in Fig. 4.17. Let the fourth force, acting at the point P with coordinates (x,y), have components X and Y parallel to the axes. Then you can write down three equations.

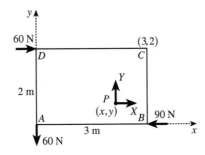

Fig. 4.17

$\mathcal{R}$ (x-direction) $X + 60 = 90$, so $X = 30$.

$\mathcal{R}$ (y-direction) $Y = 60$.

$\mathcal{M}(P)$ $60x = 90y + 60(2 - y)$, which can be simplified to $y = 2x - 4$.

Notice that there are four unknowns but only three equations. You should expect this; $y = 2x - 4$ is the equation of the line of action of the force, and it doesn't matter where along the line of action the fourth force is applied.

As a check, Fig. 4.18 shows the force with components $X = 30$ and $Y = 60$ combined to give a single resultant force. You will see that this makes an angle $\tan^{-1} 2$ with the x-direction, so it is parallel to the line of action with gradient 2.

73

The fourth force, of magnitude $30\sqrt{5}$, should be applied at some point along the line $y = 2x - 4$. The most convenient points would be where this line cuts the boundary of the plate, that is either at C with coordinates $(3,2)$, or at $(2,0)$, 2 metres from A along the edge AB.

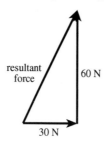

resultant force

60 N

30 N

Fig. 4.18

4.4 Indeterminate problems

Suppose that a uniform shelf AC of length 4 metres and weight W newtons rests on supports at both ends and at B, 1 metre from C. How much of the weight is taken by each support?

Let the forces at A, B and C be (in newtons) X, Y and Z (Fig. 4.19). There are three unknowns, so you need three equations. You might try

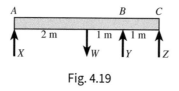

Fig. 4.19

$$\mathcal{R}(\uparrow) \quad X + Y + Z = W,$$
$$\mathcal{M}(A) \qquad 3Y + 4Z = 2W, \text{ and}$$
$$\mathcal{M}(C) \quad 4X + Y \qquad = 2W.$$

But if you try to solve these for X, Y and Z, you will be unlucky. For example, if you try to eliminate Z by taking the second equation from 4 times the first, you will get $4(X + Y + Z) - (3Y + 4Z) = 4W - 2W$, or $4X + Y = 2W$, which is the third equation over again.

The reason is that the problem does not have a unique solution. You could remove the support at B completely and just support the shelf at A and C, with $X = Z = \frac{1}{2}W$ and $Y = 0$. Or you could remove the support at C and just support the shelf at A and B, with $X = \frac{1}{3}W$, $Y = \frac{2}{3}W$ and $Z = 0$. It is easy to check that both these sets of values satisfy the three equations you started with.

A problem like this is said to be **indeterminate**. In practice, if you were to build the shelf, it would bend a little and the supports might not be exactly level; and a small adjustment in the heights would change the proportions of the weight taken by each support.

The best that you can do is to write down some inequalities satisfied by X, Y and Z. None of these forces can be negative. The equation $\mathcal{M}(A)$ above shows that, since $Z \geq 0$, $Y \leq \frac{2}{3}W$; and since $Y \geq 0$, $Z \leq \frac{1}{2}W$. The equation $\mathcal{M}(C)$ shows that, since $Y \geq 0$, $X \leq \frac{1}{2}W$. Also, from the equation $\mathcal{M}(B)$, $3X = Z + W$; since $Z \geq 0$, $X \geq \frac{1}{3}W$.

If you have done any linear programming, you will know how to show sets of inequalities like this graphically. In Fig. 4.20, X, Y and Z can take any set of values represented by points on the line joining $\left(\frac{1}{2}W, 0, \frac{1}{2}W\right)$ and $\left(\frac{1}{3}W, \frac{2}{3}W, 0\right)$.

These end points correspond to the cases where the supports at B and C are removed.

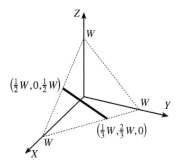

Fig. 4.20

Another example of an indeterminate problem is when you set a ladder up against a rough wall on rough ground. When, as at the beginning of this chapter, both the wall and the ground are smooth, it is impossible for the ladder to rest in equilibrium. But if there is friction at both contacts, then you have four unknowns (R, S, F and G in Fig. 4.21), and you can only write down three independent equations.

You can investigate this further for yourself in Exercise 4B Question 12.

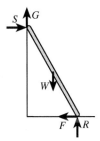

Fig. 4.21

Exercise 4B

1 A prism has a cross-section in the form of a rhombus with acute angle 40°. It is placed on a shelf which is originally horizontal, as shown in the diagram. The coefficient of friction between the prism and the shelf is 0.4. The shelf is now gradually rotated about a horizontal axis until equilibrium is broken.

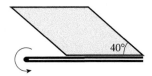

    a Find the angle through which the shelf must be rotated before the prism topples, assuming that it does not slide first.

    b Find the angle through which the shelf must be rotated before the prism slides, assuming that it does not topple first.

    Hence determine how equilibrium is broken.

2 A chair weighing 50 newtons stands on a rough floor. Its centre of mass is 30 cm behind the front legs, and 20 cm in front of the back legs. The back of the chair is 1 metre high. If a gradually increasing horizontal force is applied at the top of the back of the chair in a forward direction, the chair slides forwards. If a force of the same magnitude is applied at the same point in the opposite direction, the chair topples. What can you say about the coefficient of friction?

3 An Arctic explorer drags a sledge across a horizontal ice-field by means of a rope attached to his body harness. The rope is 5 metres long, and attached to him at a height of 1.4 metres. The sledge is 5 metres long, and the coefficient of friction between the sledge and the ice is 0.2. Where must the centre of mass of the loaded sledge be if the sledge is not to tip up as it is pulled?

4 A cylindrical tin has radius 6 cm and height 30 cm, and weighs 100 newtons. The tin stands on a table, 40 cm from the edge, which is smooth and rounded. A string is attached to the point of the top rim of the tin closest to the edge of the table; it passes over the edge and holds a bucket at its other end. Water is poured into the bucket until equilibrium is broken.

   a How much must the bucket weigh before the tin starts to topple over?

   b What can be said about the coefficient of friction between the tin and the table if the tin doesn't slide before it begins to topple?

5 A uniform cubical box of weight W stands on a horizontal floor. The coefficient of friction is μ. To try to pull the box along the floor, a force of magnitude P is applied at the centre of one of the top edges, at an angle α above the horizontal, where $0 \leq \alpha < \frac{1}{2}\pi$.

   a Show that, if the box does not topple, it will slide when $P = \dfrac{\mu W}{\cos\alpha + \mu\sin\alpha}$.

   b Show that, if the box does not slide, it will topple when $P = \dfrac{W}{2\cos\alpha}$.

   The value of P is gradually increased from zero. What condition must be satisfied by μ and α if the box begins to slide before it topples?

   Does the result remain valid if $-\frac{1}{2}\pi < \alpha < 0$?

6 A portable lamp consists of a uniform circular base with a perpendicular pole through its centre. Its centre of mass is at a height h above its base. When it is placed on a sloping ramp, it is on the point of both sliding down the ramp and toppling. A force is now applied parallel to the ramp to try to push it up the slope. Show that, if the lamp is to slide rather than topple, the force must be applied below the level of the centre of mass.

7 Forces act on the rectangular plate $ABCD$ as shown. The plate is in equilibrium. Find F and x.

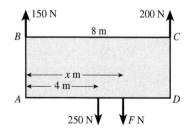

8 Forces act on the plate $ABCD$ as shown. The distance AB is 4 metres. Given that the plate is in equilibrium, find

a F, **b** the angle $\alpha°$, **c** the distance AD.

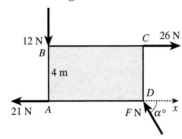

9 Forces act on the triangular plate ABO as shown. The coordinates of A and B are $(L,0)$ and $(0,L)$. Find the components of F in the x- and y-directions, and the equation of the line of action of F.

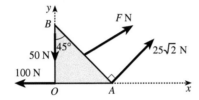

10 Some of the forces which act on a rigid object, and the points at which they are applied, are given; the units are newtons and metres. Where possible, find another force which, when combined with the given ones, will keep the object in equilibrium. Find also the equation of the line along which this force must act.

a $\begin{pmatrix} 2 \\ 0 \end{pmatrix}$ at $(3,2)$, $\begin{pmatrix} 0 \\ 4 \end{pmatrix}$ at $(1,0)$

b $\begin{pmatrix} 3 \\ -2 \end{pmatrix}$ at $(0,3)$, $\begin{pmatrix} 1 \\ 3 \end{pmatrix}$ at $(-2,1)$

c $\begin{pmatrix} 2 \\ 0 \end{pmatrix}$ at $(7,2)$, $\begin{pmatrix} 1 \\ -3 \end{pmatrix}$ at $(1,0)$, $\begin{pmatrix} -3 \\ 1 \end{pmatrix}$ at $(4,-1)$

d $\begin{pmatrix} 5 \\ -3 \end{pmatrix}$ at $(1,2)$, $\begin{pmatrix} -2 \\ 1 \end{pmatrix}$ at $(6,5)$, $\begin{pmatrix} -3 \\ 2 \end{pmatrix}$ at $(0,-1)$

e $\begin{pmatrix} 2 \\ 0 \end{pmatrix}$ at $(1,4)$, $\begin{pmatrix} 0 \\ -3 \end{pmatrix}$ at $(5,2)$, $\begin{pmatrix} 2 \\ 3 \end{pmatrix}$ at $(0,0)$

11 A uniform box has a central square cross-section $ABCD$ with sides of length 60 cm. It stands on a smooth slope at an angle of $30°$ to the horizontal, with the side DC in contact with the slope and D below C. Equilibrium is maintained by a horizontal force applied at A. If the resultant normal contact force acts at the point E, calculate the distance CE.

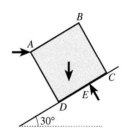

12 This question refers to Fig. 4.21. Consider the special case in which the ladder is in equilibrium with the top at a height of 4 metres above the ground and the foot at a distance of 2 metres from the wall.

a Write down equations of resolving vertically and horizontally, and moments about the top and the foot of the ladder. Show that you cannot solve these to find the four unknowns R, S, F and G.

b Write down an equation of moments about the corner where the ground meets the wall. Does this, combined with your equations in part **a**, enable you to find the four unknowns?

c Is it possible to have equilibrium with

i F,   **ii** G

acting in the direction opposite to that shown in the figure?

d If equilibrium is limiting with F and G acting in the directions shown in the figure, and if the coefficient of friction is the same at both contacts, find the coefficient of friction.

e Show that, if the coefficients of friction at the foot and the top of the ladder are μ_1 and μ_2 respectively, then $\mu_1(\mu_2 + 4) \geq 1$.

Miscellaneous exercise 4

1 Forces act on the plate $OABC$ as shown. The distance OC is 5 metres. Given that the plate is in equilibrium, find the equation of the line of action of the force of magnitude F newtons.

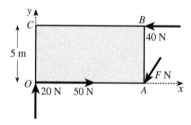

2 A non-uniform rod AB, of length 0.6 m and weight 9 N, has its centre of mass 0.4 m from A. The end A of the rod is in contact with a rough vertical wall. The rod is held in equilibrium, perpendicular to the wall, by means of a light string attached to B. The string is inclined at 30° to the horizontal. The tension in the string is T N (see diagram).

i Calculate T.

ii Find the least possible value of the coefficient of friction at A.

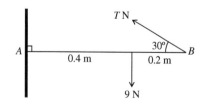

(Cambridge International AS and A Level Mathematics 9709/51 Paper 5 Q1 November 2011)

3   A uniform rod AB has weight $6\,\text{N}$ and length $0.8\,\text{m}$. The rod rests in limiting equilibrium with B in contact with a rough horizontal surface and AB inclined at $60°$ to the horizontal. Equilibrium is maintained by a force, in the vertical plane containing AB, acting at A at an angle of $45°$ to AB (see diagram). Calculate

   i   the magnitude of the force applied at A,

   ii   the least possible value of the coefficient of friction at B.

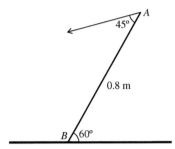

(Cambridge International AS and A Level Mathematics 9709/51 Paper 5
Q2 November 2012)

4   A non-uniform rod AB of weight $6\,\text{N}$ rests in limiting equilibrium with the end A in contact with a rough vertical wall. $AB = 1.2\,\text{m}$, the centre of mass of the rod is $0.8\,\text{m}$ from A, and the angle between AB and the downward vertical is $\theta°$. A force of magnitude $10\,\text{N}$ acting at an angle of $30°$ to the upwards vertical is applied to the rod at B (see diagram). The rod and the line of action of the $10\,\text{N}$ force lie in a vertical plane perpendicular to the wall. Calculate

   i   the value of θ,

   ii   the coefficient of friction between the rod and the wall.

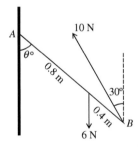

(Cambridge International AS and A Level Mathematics 9709/51 Paper 5
Q2 June 2014)

5   A uniform beam AB has length $2\,\text{m}$ and mass $10\,\text{kg}$. The beam is hinged at A to a fixed point on a vertical wall, and is held in a fixed position by a light inextensible string of length $2.4\,\text{m}$. One end of the string is attached to the beam at a point $0.7\,\text{m}$ from A. The other end of the string is attached to the wall at a point vertically above the hinge. The string is at right angles to AB. The beam carries a load of weight $300\,\text{N}$ at B (see diagram).

   i   Find the tension in the string.

The components of the force exerted by the hinge on the beam are $X\text{N}$ horizontally away from the wall and $Y\text{N}$ vertically downwards.

   ii   Find the values of X and Y.

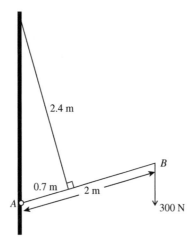

(Cambridge International AS and A Level Mathematics 9709/05 Paper 5
Q3 November 2007)

6 A uniform beam AB has length 2 m and weight 70 N. The beam is hinged at A to
a fixed point on a vertical wall, and is held in equilibrium by a light inextensible
rope. One end of the rope is attached to the wall at a point 1.7 m vertically above
the hinge. The other end of the rope is attached to the beam at a point 0.8 m
from A. The rope is at right angles to AB. The beam carries a load of weight
220 N at B (see diagram).

  i Find the tension in the rope.

  ii Find the direction of the force exerted on the beam at A.

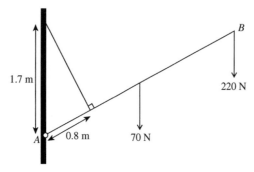

(Cambridge International AS and A Level Mathematics 9709/51 Paper 5 Q4
November 2010)

7 An L-shaped block has weight W and dimensions as shown in the
diagram. The block is at rest on a horizontal table in the position
shown. When a horizontal force P is applied to the mid-point of an
upper edge, at right angles to the vertical faces, the block topples
if the direction of the force is as shown in the first diagram, and
slides if the direction is as shown in the second diagram.

  Show that $\frac{1}{12} < \mu < \frac{5}{12}$, where μ is the coefficient of friction between
the block and the table.

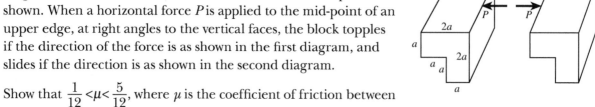

8   An L-shaped prism has weight W and dimensions as shown. The prism rests in contact with horizontal ground. A force of magnitude T is applied to the point A of the prism in a direction making an angle β with the ground, as shown in the diagram.

   a   Given that A is on the point of lifting off the ground, show that $T \sin\beta = \dfrac{23}{40}W$.

   b   Given that the coefficient of friction between the prism and the ground is 0.5, and that A is on the point of moving along the ground, show that $T(\sin\beta + 2\cos\beta) = W$.

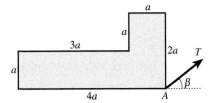

   Deduce that, if T is gradually increased from zero, the prism will slide before it starts to lift at A if $\tan\beta < \dfrac{46}{17}$.

9   A uniform sheet-metal trough of weight W has a cross-section $ABCD$ in which $AB = BC = CD$ and angle $ABC = 150° =$ angle BCD. Initially the trough is placed with AB in contact with the ground, and a small block of weight W is placed on it at the mid-point of AB. Show that the trough can remain at rest in equilibrium in this position.

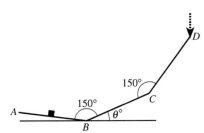

   A small downwards force is now applied at D, so that the trough rotates slowly about the edge through B. The figure shows the situation when BC makes an angle $\theta°$ with the horizontal, where $\theta < 30$. The coefficient of friction between the block and the trough is 0.4. Determine whether the block starts to slide down AB before the trough falls on its base, or vice versa.

10 The diagram shows a uniform plank of weight W in equilibrium, with its lower end on rough ground and its upper end against a rough fixed plane inclined at 60° to the horizontal. The angle between the plank and the horizontal is 30°. The normal and frictional components of the contact force exerted on the plank by the ground are R and F respectively. The normal and frictional components of the contact force exerted on the plank by the inclined plane are S and G respectively.

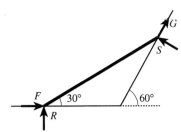

   Given that $9G = \sqrt{3}S$,

   a   find F, G, R and S in terms of W,

   b   show that $\dfrac{F}{R} = \dfrac{2}{7}\sqrt{3}$.

11 *ABCD* is a central cross-section of a uniform rectangular block of mass 35 kg. The lengths of *AB* and *BC* are 1.2 m and 0.8 m respectively. The block is held in equilibrium by a rope, one end of which is attached to the point *E* of a rough horizontal floor. The other end of the rope is attached to the block at *A*. The rope is in the same vertical plane as *ABCD*, and *EAB* is a straight line making an angle of 20° with the horizontal (see diagram).

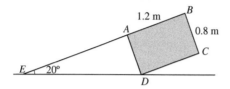

i Show that the tension in the rope is 187 N, correct to the nearest whole number.

ii The block is on the point of slipping. Find the coefficient of friction between the block and the floor.

(Cambridge International AS and A Level Mathematics 9709/05 Paper 5 Q5 November 2008)

12 The diagram shows a uniform plank of weight *W* leaning in equilibrium against a vertical wall. The angle between the plank and the horizontal is α. The normal and frictional components of the contact force exerted on the plank by the ground are *R* and *F* respectively. The normal and frictional components of the contact force exerted on the plank by the wall are *N* and *S* respectively. The coefficient of friction between the ground and the plank is μ.

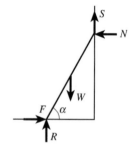

a Show that $W + 2F\tan\alpha = 2R$.

b Making the assumption that the wall is smooth, state the value of *R* in terms of *W* and show that $F = \dfrac{W}{2\tan\alpha}$. Deduce that the assumption is false if $\mu < \dfrac{1}{2\tan\alpha}$.

c Making the assumptions instead that the wall is near smooth, and that $S = kN$ where *k* is small, show that $\dfrac{F}{R} = \dfrac{1}{2\tan\alpha + k}$. Given that $\tan\alpha = 2$, that $\mu = 10k$ and that the plank is on the point of slipping, find μ.

13 Fig. I shows the cross-section of a uniform solid. The cross-section has the shape and dimensions shown. The centre of mass *C* of the solid lies in the plane of this cross-section. The distance of *C* from *DE* is *y* cm.

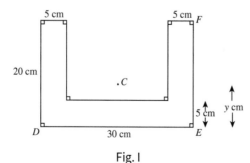

Fig. I

i Find the value of *y*.

The solid is placed on a rough plane. The coefficient of friction between the solid and the plane is μ. The plane is tilted so that *EF* lies along a line of greatest slope.

ii The solid is placed so that *F* is higher up the plane than *E* (see Fig. II). When the angle of inclination is sufficiently great the solid starts to topple (without sliding). Show that $\mu > \frac{1}{2}$.

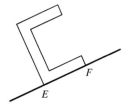

Fig. II

iii The solid is now placed so that *E* is higher up the plane than *F* (see Fig. III). When the angle of inclination is sufficiently great the solid starts to slide (without toppling). Show that $\mu < \frac{5}{6}$.

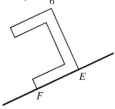

Fig. III

(Cambridge International AS and A Level Mathematics 9709/05 Paper 5 Q7 November 2007)

14 A uniform square lamina *ABCD* is in equilibrium in a vertical plane with the corner *A* in contact with rough horizontal ground and the edge *AD* inclined at angle θ to the horizontal. The lamina is held in this position by a force of magnitude *P* acting at *D* in the direction *DC*, as shown in the diagram.

The frictional force on the lamina at *A* has magnitude *F* and the normal contact force has magnitude *R*. Show that

$$\frac{F}{R} = \frac{t - t^2}{1 + t + 2t^2} \text{, where } t = \tan\theta.$$

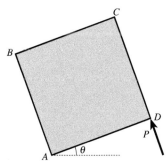

Hence show that the position of equilibrium is possible for all values of θ between 0 and $\frac{1}{4}\pi$ if the coefficient of friction between the lamina and the ground exceeds $\frac{1}{7}$.

15 A circular pipe is fixed with its axis horizontal. A uniform plank ABC, of length $2L$ and weight W, rests in a vertical plane which is perpendicular to the axis of the pipe. The plank touches the pipe at the point B, where $AB = \frac{1}{2}L$, and the end C of the plank is in contact with horizontal ground. The plank makes an angle α with the horizontal. The normal and frictional components of the contact force exerted by the ground on the plank at C are denoted by R and F respectively.

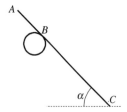

a Assuming the contact between the pipe and the plank is smooth, show that
$$\frac{F}{R} = \frac{2\tan\alpha}{3\tan^2\alpha + 1}.$$

b Given that the frictional component of the contact force at B is k times its normal component, show that $\dfrac{F}{R} = \dfrac{2(\tan\alpha - k)}{3\tan^2\alpha + 1 - 2k\tan\alpha}$.

Chapter 5
Elastic strings and springs

This chapter introduces a model to describe the behaviour of strings and springs whose length depends on the force applied to them. When you have completed it, you should

- be able to use Hooke's law as a model relating to the extension of a string, or to the extension and compression of a spring
- understand the terms 'stiffness' and 'modulus of elasticity'
- be able to derive and use the formula for the work done in stretching a string, or in stretching and compressing a spring
- understand that elastic forces are conservative, and know how to include elastic potential energy in energy calculations.

5.1 The elastic string model

So far, when dealing with mechanical systems which include strings, ropes, chains or cables, there has been no suggestion that these might stretch when pulled. In fact, in M1 Section 7.3, the model of an 'inextensible' cable was introduced, with the property that the two trucks connected by the cable move with the same speed and acceleration.

This is a very good approximation to reality in many cases, but every cable will stretch a bit when its ends are pulled apart, and some materials are made intentionally so that they stretch a lot. This doesn't just apply to substances called 'elastic', but also for example to the synthetic materials used in making ropes for rock climbers and bungee jumpers.

How does the force pulling the ends apart affect the length of a piece of elastic? You already have enough experience to know some basic facts.

- The harder you pull, the greater the length.

- If you don't pull excessively hard, the same pull will produce the same stretch each time you apply it, and the elastic will revert to its original length when you stop pulling.

- If you pull too hard, the material will be permanently distorted and lose its elastic property.

These observations now have to be converted into a mathematical model.

A few definitions are needed. The original length is called the **natural length**, usually denoted by l. The amount by which the length exceeds the natural length is the **extension**. Also, the force pulling the ends is equal to the tension. Denoting the extension by x and the tension by T, the first two points are equivalent to stating that

$T = f(x)$, where f is an increasing function and $f(0) = 0$.

The third point states that this rule only holds so long as T, and therefore x, is not too large. That is, for any piece of elastic there is a length b such that the rule holds provided that $0 \le x \le b$.

To find the form of the function f, and the value of b, you need to do precise experiments.

For materials which stretch easily, the simplest method is to fix one end, to hang weights of various magnitudes from the other end, and to measure the corresponding lengths, as in Fig. 5.1. For tougher materials more elaborate testing machinery is needed, but the principle is the same.

These experiments show that, for a wide variety of materials, the results fit pretty well a model of the form $T = kx$ for some constant k. The value of k, called the 'stiffness' or the 'elastic constant', is small for materials which stretch easily, and much larger for materials which seem to be almost inextensible.

This model is known as Hooke's law. It is named after Robert Hooke, a brilliant inventor and physical scientist who lived at about the same time as Isaac Newton.

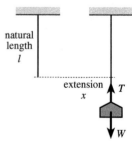

Fig. 5.1

Hooke's law **If forces are applied at the ends of a string, rope or cable, then over some range of values the tension is proportional to the extension beyond the natural length.**

But there is a drawback in using the equation $T = kx$ to describe this law. If you take two pieces of elastic, one twice as long as the other but otherwise identical, then the value of k would be half as big for the longer piece. This is because, if the same forces are applied to both, the value of T is the same; but the longer piece, which you can think of as two of the shorter pieces joined together, will stretch twice as much. Generalising this argument, you can say that k is inversely proportional to the natural length. That is $k = \dfrac{\lambda}{l}$, where λ has the same value whatever the length. This gives the usual form in which Hooke's law is used:

> **For an elastic string of natural length l, the tension T and extension x are connected by the equation $T = \dfrac{\lambda x}{l}$. The constant λ is called the** modulus of elasticity **of the string.**

Notice that, if $x = l$, then $T = \lambda$. In this case the total length of the string is $l + l = 2l$. So a physical interpretation is that the constant λ is the force that has to be applied to the ends of the string to double its length. This shows too that the modulus of elasticity can be interpreted as a force, so it is measured in the same units as force, that is in newtons.

EXAMPLE 5.1.1

A climber of mass 70 kg hangs from a rope secured at its upper end to a fixed metal ring. The natural length of the rope is 20 metres, and it stretches to 22 metres when supporting the climber. What is the modulus of elasticity of the rope?

The tension in the rope is 70×10 newtons, and the extension of the rope is 2 metres. Substitution in the equation $T = \dfrac{\lambda x}{l}$ gives

$$700 = \frac{2\lambda}{20}, \text{ so } \lambda = 7000$$

The modulus of elasticity of the rope is 7000 newtons.

EXAMPLE 5.1.2

A spider of mass 2 grams hangs from a thread which it is spinning. This has length 7 cm. A typical piece of the web has modulus of elasticity 0.05 newtons. What is the natural length of the thread from which the spider is hanging?

It is safest to convert all the data to basic SI units. The tension in the thread is 0.002×10 newtons. Let the natural length of the thread be l metres, and let the extension be x metres. Then

$$0.002 \times 10 = \frac{0.05x}{l}, \text{ and } l + x = 0.07.$$

So $x = 0.4l$, giving $l + 0.4l = 0.07$, or $l = \dfrac{0.07}{1.4} = 0.05$.

Converting back to more suitable units, the natural length of the thread is 5 cm.

EXAMPLE 5.1.3

A student uses a 40 cm length of curtain wire as a washing line. One end is attached to a hook *H* on the wall. When he has some laundry to dry, he stretches the wire so that the other end reaches another hook *K* at the same level 48 cm away. The tension in the wire is then 10 newtons. When the student hangs a wet shirt at the mid-point *M* of the wire, the wire stretches further so that *M* is 7 cm below the mid-point *N* of *HK*. How much does the wet shirt weigh?

In metre units the natural length of the wire is 0.4 m, and when the extension is 0.08 m the tension is 10 N. From this you can calculate the modulus of elasticity, using the equation $10 = \dfrac{0.08\lambda}{0.4}$ which gives $\lambda = 50$ (in newtons).

Now since the modulus of elasticity is defined so that it is independent of the length of the wire, you can use this value of λ for each of the two parts of the wire, *HM* and *MK* in Fig. 5.2.

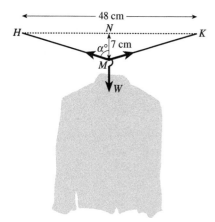

Fig. 5.2

By Pythagoras' theorem, each of these has length $\sqrt{24^2 + 7^2}$ cm = 25 cm. The natural length of each part is 0.2 m, so the extension is 0.05 m and the tension is $\dfrac{50 \times 0.05}{0.2}$ N, which is 12.5 N.

If the weight of the shirt is *W* newtons, and the angle *NMH* is $\alpha°$,

$\mathcal{R}(\uparrow)\ 2 \times 12.5 \times \cos\alpha° = W.$

Since $\cos\alpha° = \dfrac{7}{25}$, $W = 7$.

The wet shirt weighs 7 newtons.

EXAMPLE 5.1.4

Fig. 5.3 shows the design for an adjustable desk lamp in which the light bulb can rest in various positions. *O* is a point on the base, and a vertical column fixed at *O* has a small pulley *P* at the top. A rigid arm *OL* is hinged at *O* so that it can rotate in a vertical plane. A lamp is attached to the arm at *L*. An elastic cord passes over the pulley, with one end fixed at O and the other end fixed to the arm at L. The natural length of the cord is equal to the height of the column OP, so that the extension of the cord is equal to PL.

The column OP has height h, and the arm OL has length a. The arm and the lamp together have weight W, and the centre of mass is at a point G of OL such that $OG = b$. The modulus of elasticity of the cord is λ. Show that, if a certain relation holds connecting λ, W, a and b, the arm OL can rest in equilibrium at any angle to the vertical.

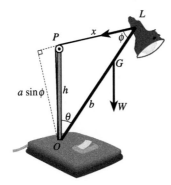

Fig. 5.3

Denote the angle POL by θ, the angle OLP by ϕ and the length PL by x.

The tension in the cord is $\dfrac{\lambda x}{h}$ so, for the arm OL to be in equilibrium,

$$M(O) \frac{\lambda x}{h} \times a \sin \phi = W \times b \sin \theta$$

This can be rearranged as

$$\frac{x}{\sin \theta} \times \lambda a = \frac{h}{\sin \phi} \times Wb$$

But by the sine rule in triangle OLP, $\dfrac{x}{\sin \theta} = \dfrac{h}{\sin \phi}$.
So the arm will be in equilibrium if $\lambda a = Wb$.

This equation doesn't involve θ, so if it is satisfied the arm will be in equilibrium at any angle to the vertical. It is interesting to notice that the equation also doesn't involve h, so the column can be made of any convenient height provided that the cord is also cut to this natural length.

5.2 Springs and rods

You will find springs in many household objects, such as door catches, mattresses, lamp holders and clocks. They come in a variety of shapes, but this section is concerned only with 'helical springs', which are in the shape of a helix (see Fig. 5.4) and which expand and contract in a direction along the axis of the helix. (A helix is the path followed if you climb up the curved surface of a cylinder at a constant angle.)

Fig. 5.4

From the point of view of mechanics, the essential difference between springs and elastic strings is that springs can be used in either tension or compression. Where a spring is used only in tension, for example in a spring balance (see M1 Example 3.4.2), it would be possible to replace the spring by a length of stiff elastic. But this would not be true of springs such as those in a mattress, which are used only in compression. Some exercise devices contain springs which are used both in tension and compression.

Hooke's law also applies to springs, but with the difference that both T and x can be either positive or negative. When x is negative, the spring is compressed and it exerts a thrust outwards at each end. So the equation $T = kx$ can be applied for values of x in an interval $-c \leq x \leq b$, where b and c are positive constants. It is more common to express the strength of a spring in terms of the stiffness k (in units such as newtons per metre) rather than the modulus of elasticity λ, although the constant λ may still appear in more theoretical examples.

A metal rod can also exert forces of either tension or compression, and will expand or contract slightly according to Hooke's law, with a very large value of k.

EXAMPLE 5.2.1

Fig. 5.5 shows the design of the bar in a toilet-roll holder, which fits between the two arms of a metal frame 134 mm wide. When out of the frame, the cylindrical part of the bar just fits between the two arms. The projections at the ends are 6 mm long, and they fit into cavities 2 mm deep in the frame. To compress the bar so as to get it out of the frame requires a force of 12 newtons. How much force does the bar exert on each arm when it is in place?

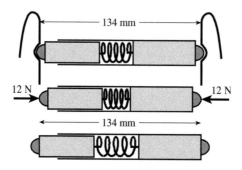

Fig. 5.5

To get the bar out of the frame its natural length must be reduced by 12 mm, the length of the two projections. This requires a force of 12 N, so the stiffness of the spring is 1 newton per millimetre.

When the bar is in place in the frame, the bar is compressed by 8 mm. The force exerted by the bar on each arm is therefore 8 newtons.

Notice that, since the spring is concealed inside the bar, its length isn't known. It is therefore impossible to determine the modulus of elasticity. But you only need to know the stiffness to answer the question.

EXAMPLE 5.2.2

Fig. 5.6 shows a design for a diving board. The board is uniform, 2 metres long and has mass 35 kg. It is hinged at one end, and is supported in a horizontal position 50 cm above the edge of the pool by a spring 1.4 metres from the hinge. It is specified that, when a boy of mass 49 kg stands at the pool end of the board, that end should not go down by more than 2 cm. Modelling the board as rigid, calculate the natural length and the modulus of elasticity of the spring needed to just satisfy that condition.

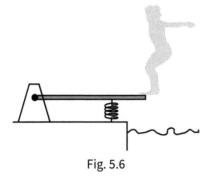

Fig. 5.6

Let the thrust from the spring be P newtons when the board is not in use, and Q newtons with the boy standing at the end of the board.

Then

$$\mathcal{M}(\text{hinge}) \quad 1.4P = 35 \times 10 \times 1 \text{ and}$$
$$1.4Q = 35 \times 10 \times 1 + 49 \times 10 \times 2,$$

which give $P = 250$ and $Q = 950$. (This is actually a slight approximation, since the board is no longer horizontal when the boy is standing on it. But the error is very small, as you can calculate if you wish.)

If the board is to go down by 2 cm at the end, it will go down by 1.4 cm at the point of support from the spring.

Let the stiffness of the spring be k newtons per metre, and compression of the string be x m when the board is not in use.

Then $P = kx$ and $Q = k(x + 0.014)$.

Subtracting these gives

$Q - P = 0.014\,k.$

But $Q - P = 950 - 250 - 700$, so $k - 50\,000$.

Substituting this and $P = 250$ into $P = kx$ then given $x = 0.005$.

The natural length of the spring is therefore 0.505 metres, and the modulus of elasticity is $50\,000 \times 0.505$ newtons, which is 25 250 newtons.

Therefore a spring of natural length 50.5 cm and modulus of elasticity 25.25 kN will be needed.

Exercise 5A

1 A cable 20 metres long is being used to raise a container of mass 25 tonnes. After the cable has been pulled taut, the upper end has to move a further 1.6 cm before the container lifts clear of the ground. Calculate the modulus of elasticity of the cable.

2 A ring of mass 20 grams is lifted gently off a table at a steady speed by an elastic thread of natural length 28 cm and modulus of elasticity 7 newtons. What is the length of the thread while it is lifting the ring?

3 In an 8-a-side tug-of-war the front men in each team are 10 metres apart, and the men behind them are spaced out at intervals of 2 metres. The rope has modulus of elasticity 50 000 newtons. On the command 'take the strain' each competitor pulls on the rope with a force of 120 newtons. How much does the rope stretch

a between the two front men,

b between the two anchor men at the ends?

4 A box of weight 20 newtons is placed on a table. It is to be pulled along by an elastic string with natural length 15 cm and modulus of elasticity 5 newtons. The coefficient of friction between the box and the table is 0.4. Holding the string horizontally by its loose end, and beginning with the string just taut, how far would you have to pull before the box starts to move?

5 A mattress is 18 cm deep. When a 75 kg sleeper lies on it, 15 of the springs are compressed by an average of 0.5 cm. Calculate the force needed to compress each spring by 1 cm.

6 A new bulb with a bayonet fitting is to be inserted into a vertical lamp-holder. The mass of the bulb is 40 grams. To insert the bulb it is first placed in the holder so that it rests on the two vertical springs. It then has to be pushed down 6 mm against the springs and twisted; then, when it is released, it rises 2 mm and is held firm by the pins which fit into the slots in the holder. The maximum force you need to exert during the process is 2 newtons. Find the force holding the bulb in position.

7 A uniform rod AB of weight W and length l is hinged to a wall at A and has a small ring at B. An elastic string of natural length l and modulus of elasticity λ has one end tied to the hinge at A, passes through the ring at B and is then attached to a point C on the wall at a distance c vertically above A. Show that the rod can rest horizontally in equilibrium provided that $\lambda c = \frac{1}{6}lw$

8 An elastic string of natural length l and modulus of elasticity λ has its two ends attached to pegs a distance a apart. It is pulled aside by a force applied at the mid-point of the string at right angles to the line of the pegs. Find the force necessary to hold it in position when the two parts of the string and the line joining the pegs form an equilateral triangle.

9 An elastic band is made of material with modulus of elasticity 2 newtons. The natural length of the band is 12 cm, and a lucky mascot of weight 10 newtons hangs from it. The band is supported in two ways:

a from a single peg,

b looped over two pegs at the same level 12 cm apart.

In each case, find how far the point of attachment of the mascot is below the peg(s) when it is in equilibrium.

10 The ends of an elastic string of natural length $2l$ and modulus λ are attached to two pegs a distance $4l$ apart. The string is held firmly at its mid-point H, and this point is then moved along the line between the pegs. Find expressions for the force needed to hold it steady at a distance x from its original position

a when $x < l$,   **b** when $l < x < 3l$,   **c** when $x > 3l$.

Illustrate your answer with a graph.

11   Two elastic strings, each of natural length l and modulus of elasticity λ, are clamped together at their ends to form a single string of natural length l. What is its modulus of elasticity?

12   A rod is hinged at one end and is kept horizontally in equilibrium by a vertical elastic string of natural length l attached to it at a distance x from the hinge. If the length of the string is then y, find the form of the equation connecting x and y. Use this to show on a diagram the locus of the upper end of the string as x is varied.

13   A point A of a ceiling is directly above a point B of the floor 2.5 metres below. A and B are joined by a light spring of natural length 2 metres; the tension in the spring is then 20 newtons. A lump of putty of weight 4 newtons is now attached to the spring at a point 1 metre above the floor. At what height above the floor will the putty rest in equilibrium?

5.3   Work done in stretching and compressing

When you stretch an elastic string, you are not just concerned to provide the force needed to hold it steady once it is stretched. You also have to stretch it to that length in the first place, and to do that you have to do work.

In M1 Section 8.1 the work done on an object by a constant force F acting along a line was defined as Fx, where x is the displacement of the object along the line. This can be represented in a force–displacement graph such as Fig. 5.7 by the shaded area under the graph.

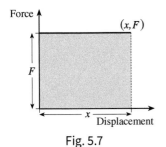

Fig. 5.7

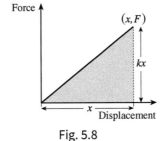

Fig. 5.8

You can't use the formula Fx for the work done as you stretch a string, because the force varies. The force–displacement graph has equation $F = kx$, and is therefore a line segment through the origin with gradient k, as in Fig. 5.8. However, the work done by the force is still represented by the area under the graph. This is calculated as the area of the shaded triangle, which is $\frac{1}{2}x \times (kx)$, that is $\frac{1}{2}kx^2$. Since $k = \frac{\lambda}{l}$, in terms of the modulus of elasticity this is $\frac{\lambda x^2}{2l}$.

Since kx is the tension when the string is stretched to extension x, this formula can be written as $\frac{1}{2} \times$ final tension $\times$ extension. It sometimes happens that you know the tension and the extension, but have not calculated the stiffness, and this is then the simplest way of finding the work done.

The formula also holds when you compress a spring. Although F and x are then both negative, $\frac{1}{2}kx^2$ is positive. Whether you are stretching or compressing a spring, you have to do a positive amount of work.

93

When a string is stretched, or a spring is stretched or compressed, starting at its natural length l, the work done in changing its length by an amount x is

$$\frac{1}{2}kx^2 = \frac{\lambda x^2}{2l}$$

where k is the stiffness and λ is the modulus of elasticity. This is $\frac{1}{2} \times$ **final tension $\times$ extension**, or $\frac{1}{2} \times$ **final thrust $\times$ reduction in length**.

EXAMPLE 5.3.1

In Example 5.1.3, how much work does the student do in fixing the curtain wire to the second hook?

To get the answer in joules, it is important to begin by converting the lengths from centimetres to metres. The work can then be calculated as

$$\frac{\lambda x^2}{2l} = \frac{50 \times 0.08^2}{2 \times 0.4} = 0.4$$

or as $\frac{1}{2} \times$ final tension $\times$ extension $= \frac{1}{2} \times 10 \times 0.8 = 0.4$.

The work done in fixing the wire to the second hook is 0.4 joules.

5.4 Elastic potential energy

What have a bow-and-arrow and a wind-up clock in common? Both use the elastic properties of materials as a source of energy. An archer does work in flexing the bow, and the energy thus created is converted into the kinetic energy of the arrow. Winding up a clock stores energy in the mainspring, which is then used to operate the time-keeping mechanism.

The same principle holds when you stretch or compress a helical spring, or stretch an elastic string. Elastic forces, like gravity, are conservative. This means that the work you do in deforming the spring or string is not wasted, but stored up as 'elastic potential energy' for possible use later. The formulae in Section 5.3 can then be restated as follows:

A string or spring, stretched or compressed by a distance x, possesses elastic potential energy E of amount $E = \frac{1}{2}kx^2 = \frac{\lambda x^2}{2l}$.

When you have a mechanical system that includes elastic materials, elastic potential energy has to be included alongside kinetic energy and gravitational potential energy in the total energy equation. The conservation of energy principle (see M1 Chapter 9) then becomes:

If no work is done by non-conservative forces, the total energy (kinetic and potential, both gravitational and elastic) remains constant.

EXAMPLE 5.4.1

In a toy gun a cork of mass 4 grams is shot out of the barrel by the release of a spring, which is compressed through a distance of 5 cm. A force of $6x$ newtons is needed to keep the spring compressed by x cm. Find the speed with which the cork leaves the barrel.

Converting the data to basic SI units, the mass of the cork is 0.004 kg and the compression of the spring is 0.05 m. The stiffness k, of 6 N cm⁻¹, is 600 N m⁻¹. The potential energy of the compressed spring is therefore $\frac{1}{2} \times 600 \times 0.05^2$ joules, and this is converted into kinetic energy of $\frac{1}{2} \times 0.004 \times v^2$ joules, where v m s⁻¹ is the speed of the cork. Assuming that the barrel is smooth, and that the gun is held still as it is fired, there are no other forces which do work. So

$$0.75 = 0.002v^2, \text{ which gives } v = 19.36....$$

The cork leaves the barrel at a speed of 19.4 m s⁻¹.

EXAMPLE 5.4.2

One end of an elastic string, of natural length 1.5 metres and modulus of elasticity 50 newtons, is attached to a hook in the ceiling. A particle of mass 2 kg is attached at the other end, and hangs in equilibrium, as shown in Fig. 5.9. The particle is then pulled down a distance of b metres, and released. Find how high it rises
a if $b = 0.4$,   **b** if $b = 0.9$.

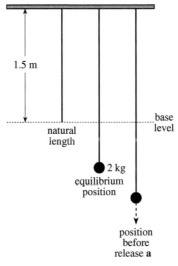

Fig. 5.9

If the extension of the string is c metres when the particle is in equilibrium, the tension in the string is equal to the weight, so

$$\mathcal{R}(\uparrow) \frac{50c}{1.5} = 2 \times 10 \text{ which gives } c = 0.6.$$

a   If $b = 0.4$, the extension of the string is 1 m when the particle is released. There is then no kinetic energy, so the total energy is the sum of the gravitational and the elastic potential energy.

You need to choose a base level from which to measure the gravitational energy, and a convenient choice is 1.5 m below the hook, which is the level of the bottom of the string if there is no particle attached to it. Then the gravitational energy is $-2 \times 10 \times 1$ J and the elastic energy is $\dfrac{50 \times 1^2}{2 \times 1.5}$ J. The total energy is then $\left(-20 + \frac{50}{3}\right)$ J, which can be simplified to $-\frac{10}{3}$ J.

Now there is a problem. It is not obvious whether, at the top of its path, the particle rises above the base level or not. You can settle this by thinking how much energy there would be at that level if the particle rises that far. There would be no gravitational energy, and also no elastic energy (because the string would be unstretched in that position). So the only possible energy is kinetic, which must be positive. But since you already know that the total energy is negative, that is impossible. It follows that the string is still stretched at the top of the path.

So let the extension be x metres where the particle comes to rest at its highest point. There is then no kinetic energy, and the total energy is made up of gravitational energy of $-2 \times 10 \times x$ J and elastic energy of $\dfrac{50 \times x^2}{2 \times 1.5}$ J, making a total of $\left(-20x + \frac{50}{3}x^2\right)$ J.

Equating this to the energy when the particle is released,

$$-\tfrac{10}{3} = -20x + \tfrac{50}{3}x^2 .$$

This can be written as $5x^2 - 6x + 1 = 0$, which factorises as $(5x - 1)(x - 1) = 0$, so $x = 0.2$ or $x = 1$.

You might be surprised to get a quadratic equation with two roots, but of course $x = 1$ corresponds to the point where the particle was released, when there is also no kinetic energy. So the root required is $x = 0.2$.

The particle rises by $(1 - 0.2)$ metres, which is 0.8 metres.

b If $b = 0.9$, the extension is 1.5 m, so the gravitational energy when the particle is released is $-2 \times 10 \times 1.5$ J, and the elastic energy is $\dfrac{50 \times 1.5^2}{2 \times 1.5}$ J.

The total energy is therefore $(-30 + 37.5)$ J, which can be simplified to 7.5 J.

Because this is positive, the particle now still has positive energy at the base level where the string is unstretched; since both gravitational and elastic energy are zero at this level, this is kinetic energy. At the highest point therefore the string is unstretched, and there is no elastic energy. If the top of the path is y metres above the base level, the gravitational potential energy is $2 \times 10 \times y$ J, which is $20y$ J.

The energy equation now has the form $7.5 = 20y$ with solution $y = 0.375$.

The particle rises by $(1.5 + 0.375)$ metres, which is 1.875 metres.

In Example 5.4.2, it is important to notice that, if the string is replaced by a spring, then the answer to part **a** would be the same, but the answer to part **b** would be different. This is because, once the particle rises above the base level, the spring would be compressed, and there would be elastic energy of $\dfrac{50 \times y^2}{2 \times 1.5}$ J to add to the gravitational energy. You can check for yourself that this would lead to the quadratic equation $20y^2 + 24y - 9 = 0$ and the relevant solution is $y = 0.3$.

The particle would therefore rise by $(1.5 + 0.3)$ metres, which is 1.8 metres.

Exercise 5B

1 A boy of weight 294 N falls 2.5 metres onto a trampoline that brings him to rest after he has descended a further distance of 0.2 metres. Calculate the kinetic energy of the boy when he reaches the trampoline, and the energy stored in the trampoline when the boy is at rest.

2 A bungee jumper of mass 65 kg leaps from Kawarau Suspension Bridge 43 metres above the river. The elastic rope to which she is attached is 25 metres long. Calculate the speed of the jumper when the free-fall part of her descent is complete. Her descent ends when she just reaches the water. Calculate the energy stored in the rope at this instant.

3 An elastic string has natural length 2 metres and modulus of elasticity 45 newtons. One end is fastened to a fixed point O on a smooth table. A particle of mass 0.1 kg is attached to the other end. The particle is placed on the table 2.5 metres from O, with the string stretched, and then released. How fast is the particle moving when the string becomes slack?

   How far is the particle from O when it next comes to rest? Describe the motion after that.

4 Repeat Question 3 with the elastic string replaced by a spring with the same natural length and modulus of elasticity.

5 The ceiling of a room is 2 metres above the floor. A ball of mass m kg hangs from an elastic string attached to the ceiling. The natural length of the string is 0.8 metres, and in equilibrium the ball rests 1 metre below the ceiling. The ball is now pulled down and placed on the floor with the string stretched. Find the speed with which the ball hits the ceiling after it is released.

6 A piston of mass 0.5 kg moves in a horizontal cylinder which is closed at one end. The piston is pushed along the cylinder against the thrust from a spring which is attached to the base of the cylinder at its closed end. A force of $1500x$ newtons is required to compress the spring by x metres. The motion of the piston is opposed by a frictional force of 10 newtons from the walls of the cylinder. In the initial position the spring is unstressed, and the piston just touches the spring. The piston is then pushed through a distance of 6 cm and released.

   a How much work is done in pushing the piston?

   b With what speed does the piston emerge from the open end of the cylinder?

7 A particle of mass m hangs in equilibrium from a fixed point by an elastic string, which stretches a distance x under the weight of the particle. If you place your hand underneath the particle and raise it gently until the string is unstretched, how much work do you do?

8 A bungee jumper of mass 80 kg is attached to a bridge by a rope 30 metres long. The river runs 46 metres below the bridge. If the jump is designed to bring him to rest 1 metre above the water level, what should be the modulus of elasticity of the rope?

9 A construction worker of mass 80 kg is wearing a safety harness with modulus of elasticity 4000 newtons. He loses his footing, and when he has fallen 2 metres the harness becomes taut. How much further does he fall before he is brought to rest?

10 An elastic string of natural length 2 metres and modulus 40 newtons is stretched between two fixed pegs 4 metres apart on a smooth horizontal table. A particle of mass 0.1 kg is attached to the string at its mid-point, and pulled aside through a distance of 1 metre at right angles to the line joining the pegs. After the particle is released, how fast is it moving when it crosses the line joining the pegs?

11 The prongs of a catapult are 18 cm apart. An elastic string of unstretched length 15 cm is attached to each prong, and the strings are joined together by a small leather pouch in which the stone is placed. The pouch is pulled back horizontally, symmetrically between the prongs, until it is 40 cm behind the vertical plane of the prongs, and then let go. If the force holding the pouch in position is 30 newtons, and the mass of the stone is 50 grams, find the speed with which the stone is projected when the strings become slack.

12 In a circus act an elastic rope of natural length 8 metres is stretched between two fixed points 10 metres apart at the same level. An acrobat, of mass 50 kg, grabs the rope at its mid-point and descends vertically. She comes to rest when she has dropped 12 metres. Find the modulus of elasticity of the rope, and how fast she is moving when she has dropped 6 metres.

13 A particle of mass m is attached to an elastic string of natural length l and modulus of elasticity λ. The upper end of the string is fixed, and the particle is falling with speed u when the string becomes taut.

   a Find an expression for the speed when the string has extension x.

   b Find how far the string stretches before it comes to rest.

   c Show that the speed is greatest when the particle passes through the equilibrium position. How could this be predicted without doing any calculation?

14 An elastic rope of natural length l hangs from a tree branch. A monkey of mass m grasps the rope at its lower end, and rests in equilibrium at a distance $l + a$ below the branch. How much work does the monkey do if he climbs slowly up the rope to the branch?

Miscellaneous exercise 5

1 A particle P of mass m kg is attached to the mid-point of a light elastic string of natural length 0.8 m and modulus of elasticity 8 N. One end of the string is attached to a fixed point A and the other end is attached to a fixed point B which is 2 m vertically below A. When the particle is in equilibrium the distance AP is 1.1 m (see diagram on the right). Find the value of m.

(Cambridge International AS and A Level Mathematics 9709/05 Paper 5
Q1 June 2005)

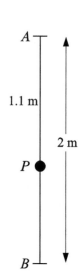

2

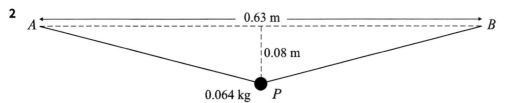

A light elastic string has natural length 0.6 m and modulus of elasticity λ N. The ends of the string are attached to fixed points A and B, which are at the same horizontal level and 0.63 m apart. A particle P of mass 0.064 kg is attached to the mid-point of the string and hangs in equilibrium at a point 0.08 m below AB (see diagram). Find

i the tension in the string,

ii the value of λ.

> (Cambridge International AS and A Level Mathematics 9709/05 Paper 5
> Q1 June 2006)

3 One end of a light elastic string, of natural length 0.5 m and modulus of elasticity 140 N, is attached to a fixed point O. A particle of mass 0.8 kg is attached to the other end of the string. The particle is released from rest at O. By considering the energy of the system, find

i the speed of the particle when the extension of the string is 0.1 m,

ii the extension of the string when the particle is at its lowest point.

> (Cambridge International AS and A Level Mathematics 9709/05 Paper 5
> Q5 June 2007)

4 A particle P of mass 0.4 kg is attached to one end of a light elastic string of natural length 1.2 m and modulus of elasticity 19.2 N. The other end of the string is attached to a fixed point A. The particle P is released from rest at the point 2.7 m vertically above A. Calculate

i the initial acceleration of P,

ii the speed of P when it reaches A.

> (Cambridge International AS and A Level Mathematics 9709/51 Paper 5
> Q2 June 2013)

5 A light elastic string has natural length 0.8 m and modulus of elasticity 16 N. One end of the string is attached to a fixed point O, and a particle P of mass 0.4 kg is attached to the other end of the string. The particle P hangs in equilibrium vertically below O.

i Show that the extension of the string is 0.2 m.

P is projected vertically downwards from the equilibrium position. P first comes to instantaneous rest at the point where $OP = 1.4$ m.

ii Calculate the speed at which P is projected.

iii Find the speed of P at the first instant when the string subsequently becomes slack.

> (Cambridge International AS and A Level Mathematics 9709/51 Paper 5
> Q3 June 2014)

6   One end of a light elastic string of natural length 1.6 m and modulus of elasticity 28 N is attached to a fixed point O. The other end of the string is attached to a particle P of mass 0.35 kg which hangs in equilibrium vertically below O. The particle P is projected vertically upwards from the equilibrium position with speed 1.8 m s^{-1}. Calculate the speed of P at the instant the string first becomes slack.

(Cambridge International AS and A Level Mathematics 9709/51 Paper 5
Q3 November 2014)

7   A light elastic string has natural length 2.4 m and modulus of elasticity 21 N. A particle P of mass m kg is attached to the mid-point of the string. The ends of the string are attached to fixed points A and B which are 2.4 m apart at the same horizontal level. P is projected vertically upwards with velocity 12 m s^{-1} from the mid-point of AB. In the subsequent motion P is at instantaneous rest at a point 1.6 m above AB.

   i   Find m.

   ii Calculate the acceleration of P when it first passes through a point 0.5 m below AB.

(Cambridge International AS and A Level Mathematics 9709/51 Paper 5
Q4 June 2012)

8   A particle P of mass 0.28 kg is attached to the mid-point of a light elastic string of natural length 4 m. The ends of the string are attached to fixed points A and B which are at the same horizontal level and 4.8 m apart. P is released from rest at the mid-point of AB. In the subsequent motion, the acceleration of P is zero when P is at a distance 0.7 m below AB.

   i   Show that the modulus of elasticity of the string is 20 N.

   ii Calculate the maximum speed of P.

(Cambridge International AS and A Level Mathematics 9709/51 Paper 5
Q5 November 2010)

9   One end of a light elastic string of natural length 1.25 m and modulus of elasticity 20 N is attached to a fixed point O. A particle P of mass 0.5 kg is attached to the other end of the string. P is held at rest at O and then released. When the extension of the string is x m the speed of P is v m s^{-1}.

   i   Show that $v^2 = -32x^2 + 20x + 25$.

   ii Find the maximum speed of P.

   iii Find the acceleration of P when it is at its lowest point.

(Cambridge International AS and A Level Mathematics 9709/05 Paper 5
Q6 June 2008)

10 A particle P of mass 1.6 kg is attached to one end of each of two light elastic strings. The other ends of the strings are attached to fixed points A and B which are 2 m apart on a smooth horizontal table. The string attached to A has natural length 0.25 m and modulus of elasticity 4 N, and the string attached to B has natural length 0.25 m and modulus of elasticity 8 N. The particle is held at the mid-point M of AB (see diagram).

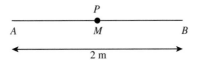

i Find the tensions in the strings.

ii Show that the total elastic potential energy in the two strings is 13.5 J.

P is released from rest and in the subsequent motion both strings remain taut. The displacement of P from M is denoted by x m. Find

iii the initial acceleration of P,

iv the non-zero value of x at which the speed of P is zero.

(Cambridge International AS and A Level Mathematics 9709/05 Paper 5
Q6 June 2009)

11 A particle P of mass 0.35 kg is attached to the mid-point of a light elastic string of natural length 4 m. The ends of the string are attached to fixed points A and B which are 4.8 m apart at the same horizontal level. P hangs in equilibrium at a point 0.7 m vertically below the mid-point M of AB (see diagram).

i Find the tension in the string and hence show that the modulus of elasticity of the string is 25 N.

P is now held at rest at a point 1.8 m vertically below M, and is then released.

ii Find the speed with which P passes through M.

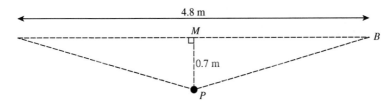

(Cambridge International AS and A Level Mathematics 9709/51 Paper 5
Q6 June 2010)

Revision exercise 1

1 A particle P of mass 0.4 kg is attached to one end of a light elastic string of natural length 1.5 m and modulus of elasticity 6 N. The other end of the string is attached to a fixed point O on a rough horizontal table. P is released from rest at a point on the table 3.5 m from O. The speed of P at the instant the string becomes slack is 6 ms^{-1}. Find

   i the work done against friction during the period from the release of P until the string becomes slack,

   ii the coefficient of friction between P and the table.

(Cambridge International AS and A Level Mathematics 9709/05 Paper 5 Q4 June 2005)

2 Two parallel rails are a horizontal distance 2 metres apart. One is slightly higher than the other, so that a girder AB of length 10 metres can be placed horizontally to pass between the rails. The end A projects 1 metre beyond the higher rail, and the end B projects 7 metres beyond the lower rail. The girder is uniform and has weight 9000 newtons.

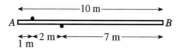

   a What is the largest weight that can be hung from the girder at any point of its length without disturbing the equilibrium?

   b If this weight is hung from B, what are the contact forces between the girder and the rails?

3 In the figure the curve AB is a snow slope mounted on a frame used for ski-jumping practice. Skiers start from rest at A, 30 metres above the horizontal ground, and launch themselves into the air from B, 10 metres above the ground, at 10° to the horizontal. C is the point on the ground vertically below B. It is calculated that, if friction, any force exerted by the skier and forces on the skier from the air are neglected, the skier will land at D. Calculate the distance CD.

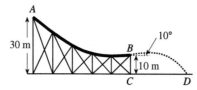

4 A sheet of metal is cut into the T-shape shown in the figure. It is supported on rails at A and B. What can you say about l if the shape can rest vertically and stably in the position shown?

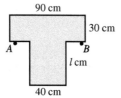

5 A light elastic string has natural length 4 m and modulus of elasticity 2 N. One end of the string is attached to a fixed point O of a smooth plane which is inclined at $30°$ to the horizontal. The other end of the string is attached to a particle P of mass 0.1 kg. P is held at rest at O and then released. The speed of P is $v\,\text{m s}^{-1}$ when the extension of the string is x m.

i Show that $v^2 = 45 - 5(x - 1)^2$.

Hence find

ii the distance of P from O when P is at its lowest point,

iii the maximum speed of P.

> (Cambridge International AS and A Level Mathematics 9709/05 Paper 5 Q6
> November 2008)

6 A flexible chain of mass m and length l hangs over a smooth horizontal nail with equal lengths on either side. Equilibrium is disturbed by pulling one end of the chain down by a very small amount, so that the chain begins to slip over the nail. Taking the level of the nail as the base-line, find the potential energy of the chain

a when it is hanging in equilibrium,

b when the last link of the chain has just slipped over the nail.

Hence find the speed with which the chain is moving vertically as the end slips off the nail.

7 The student living in room number 6 cuts this shape out of a sheet of cardboard and fixes it to the door with a drawing pin.

a How many squares away from the left upright should the pin be placed so that the numeral hangs true?

b If the student puts the pin in at the point P, how many degrees out of true will the numeral hang?

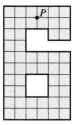

8 A light elastic string has natural length 3 m and modulus of elasticity 45 N. A particle P of weight 6 N is attached to the mid-point of the string. The ends of the string are attached to fixed points A and B which lie in the same vertical line with A above B and $AB = 4$ m. The particle P is released from rest at the point 1.5 m vertically below A.

i Calculate the distance P moves after its release before first coming to instantaneous rest at a point vertically above B. (You may assume that at this point the part of the string joining P to B is slack.)

ii Show that the greatest speed of P occurs when it is 2.1 m below A, and calculate this greatest speed.

iii Calculate the greatest magnitude of the acceleration of P.

> (Cambridge International AS and A Level Mathematics 9709/51 Paper 5 Q7
> November 2012)

9 A particle is projected from a point O on horizontal ground. The velocity of projection has magnitude $20\,\mathrm{m\,s^{-1}}$ and direction upwards at an angle θ to the horizontal. The particle passes through the point which is $7\,\mathrm{m}$ above the ground and $16\,\mathrm{m}$ horizontally from O, and hits the ground at the point A.

i Using the equation of the particle's trajectory and the identity $\sec^2\theta = 1+\tan^2\theta$, show that the possible values of $\tan\theta$ are $\frac{3}{4}$ and $\alpha = \frac{1}{2}\pi$.

ii Find the distance OA for each of the two possible values of $\tan\theta$.

iii Sketch in the same diagram the two possible trajectories.

(Cambridge International AS and A Level Mathematics 9709/51 Paper 5 Q5 June 2010)

10 A child makes a pile of four cubes on the table, with edges of length $8\,\mathrm{cm}$, $6\,\mathrm{cm}$, $4\,\mathrm{cm}$ and $2\,\mathrm{cm}$. Find the height of the centre of mass of the pile above the table

a if the cubes are solid and all made of the same material,

b if the cubes are hollow and made of the same sheet material.

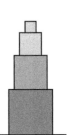

11 A ruler 1 metre long weighs 2 newtons. A boy supports it horizontally using the index fingers of his two hands. His left and right fingers are at the $0\,\mathrm{cm}$ and $80\,\mathrm{cm}$ marks respectively. He then tries to bring his fingers together slowly, applying horizontal forces of equal magnitude just large enough for one finger to slip. The coefficient of friction between each finger and the ruler is 0.5. Which finger will slip, and what initial force will be needed? What will happen after that?

Suppose now that the coefficient of friction is 0.5 if there is no movement, but that it is only 0.4 when there is movement between the surfaces. Show that in this case the first finger to slip will continue to do so until it reaches the $26\,\mathrm{cm}$ mark, and that after that the other finger will start to slip. Up to what mark will the second finger move?

12 The figure shows a racing dinghy crewed by two people each of weight 800 newtons, sailing straight ahead at constant speed. The wind on the sails is horizontal, and has a sideways component of S newtons and a forward component of F newtons, acting at a height 3.2 metres above sea level. The water exerts a sideways force on the hull and the keel of P newtons, acting 0.8 metres below sea level. To keep the boat on an even keel, the crew place their feet on the deck rail and lean out from the side of the dinghy, so that their weight acts along a line 2 metres from the centre line of the boat. Calculate S.

If the resultant wind force is at 65° to the direction of the boat, calculate the combined resistance of the air and the water to the forward motion of the boat.

13 A particle *P* is projected from a point *O* on horizontal ground. 0.4 s after the instant of projection, *P* is 5 m above the ground and a horizontal distance of 12 m from *O*.

  i Calculate the initial speed and the angle of projection of *P*.

  ii Find the direction of motion of the particle 0.4 s after the instant of projection.

    (Cambridge International AS and A Level Mathematics 9709/51 Paper 5 Q6 June 2011)

14 An earth-moving vehicle runs on two continuous metal tracks. The length of track in contact with the ground is 3 metres, and the outsides of the two tracks are 2 metres apart. The centre of the rectangular patch of ground between the tracks is *O*. The vehicle is in two parts. The part which includes the tracks and the engine has mass 2 tonnes, and its centre of mass is 0.4 metres above *O*. The other part, which includes the cabin, the earth-moving mechanism and the load, can have mass up to 3 tonnes; this part can rotate about a vertical axis above *O*. Its centre of mass may be as much as 2.4 metres above the ground, and 0.5 metres from the axis of rotation.

  a What can you say about the position of the centre of mass of the whole vehicle including its load?

  b What is the greatest angle of the slope on which the vehicle should be operated

    i with the tracks facing up or down the slope,   **ii** with the tracks facing across the slope?

15 *ABCD* is the cross-section through the centre of mass of a uniform rectangular block of weight 260 N. The lengths *AB* and *BC* are 1.5 m and 0.8 m respectively. The block rests in equilibrium with the point *D* on a rough horizontal floor. Equilibrium is maintained by a light rope attached to the point *A* on the block and the point *E* on the floor. The points *E*, *A* and *B* lie in a straight line inclined at 30° to the horizontal (see diagram overleaf).

  i By taking moments about *D*, show that the tension in the rope is 146 N, correct to 3 significant figures.

  ii Given that the block is in limiting equilibrium, calculate the coefficient of friction between the block and the floor.

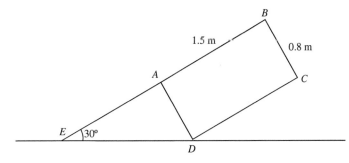

    (Cambridge International AS and A Level Mathematics 9709/51 Paper 5 Q6 November 2013)

16 The square $ABCD$ in the diagram is the central cross-section of a cuboidal block of weight W. The block stands on the ground with AD against a vertical wall. The edge through A perpendicular to $ABCD$ is smoothly hinged to the wall. A cable is attached to the block at B, and this is used to raise the block slowly off the ground until AB is vertical. The cable passes over a pulley at the point E above A such that $AE = AB$, and the other end is wound onto a drum powered by a motor. Show that, when AB makes an angle $\theta°$ with the horizontal, the tension in the cable is

$$\tfrac{1}{2}\sqrt{2W}\,\frac{\sin(45+\theta)°}{\sin(45+\tfrac{1}{2}\theta)°}.$$

Sketch graphs of $\sin(45+\theta)°$ and $0 < \theta < \tfrac{1}{2}\pi$ for values of θ from 0 to 90, and use these to describe how the tension in the cable varies as the block is raised.

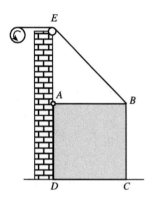

Chapter 6
Motion in a circle

This chapter deals with the connection between force, velocity and acceleration for an object modelled as a particle moving round a circular path. When you have completed it, you should

■ know the meaning of angular speed
■ know the equation connecting tangential speed and angular speed
■ know how to calculate the acceleration in circular motion with constant speed
■ be able to solve problems in two and three dimensions involving objects moving round a circular path with constant speed.

6.1 Two practical examples

Fast trains from London to Cambridge travel for some distance along a straight track at constant speed. As a passenger you are hardly aware of how fast the train is moving. By Newton's first law, the forces from the seat supporting you are the same as they are when the train is stationary.

As the train approaches Hitchin junction (see Fig. 6.1), you become conscious that it is slowing down. If you are facing backwards, the force from the back of the seat increases, so that you decelerate at the same rate as the train. Once the speed has dropped sufficiently for the train to negotiate the bend ahead, this force reverts to its previous value.

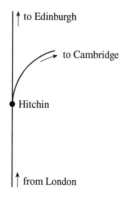

Fig. 6.1

The next force that you experience is a sideways force, and then you know that the train has left the main line and has started to round the bend on the Cambridge line. This force lasts as long as the bend continues.

Where there is force, there must be acceleration. An object moving in a curved path has an acceleration directed inwards along the normal to the curve. If the path is circular, this normal is along the radius, that is towards the centre of the circle.

You may think it surprising that an object can be accelerating if its speed is constant. But although the speed stays the same, the velocity is changing. Force is needed to change the direction of the velocity.

It was similar reasoning that led Newton to his theory of gravitational attraction. He observed that the moon goes round the earth in a nearly circular path. If there were no force, then the moon should simply move in a straight line. Since in fact it moves in a circle, there must be a force on the moon directed along the radius, that is towards the earth. This force produces an acceleration, called the gravitational acceleration.

The next step is to find how to calculate this acceleration. This is the subject of the next three sections.

6.2 Angular speed

For a particle moving along a straight line, speed is calculated as the rate at which the distance is changing. You can measure the speed of a particle moving in a circle in the same way. But often it is more convenient to measure the speed by finding the rate at which the radius is turning.

Fig. 6.2 shows a particle P moving at constant speed v round a circle with centre O and radius r. Suppose that, at a time t after it passes a point A on the circle, it has moved a distance s, and that angle AOP is θ radians. Then the speed of the particle is $\dfrac{s}{t}$, and $s = r\theta$. So the speed is $\dfrac{r\theta}{t}$, which can be rearranged as $r \times \dfrac{\theta}{t}$. The quantity which appears in this expression is called the **angular speed** of the particle about O. It is usually denoted by the greek letter ω (omega). In a diagram angular speed is shown by an arrow with a curved shaft and the same head as the velocity arrow, as in Fig. 6.2.

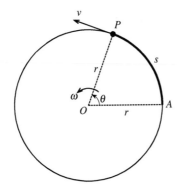

Fig. 6.2

The tangential speed of the particle can then be expressed in terms of r and ω.

A particle moving round a circle of radius r with angular speed ω has tangential speed v given by $v = r\omega$.

In SI units, with r in metres and v in $\mathrm{m\,s^{-1}}$, ω is in $\mathrm{rad\,s^{-1}}$. If it is more convenient, other units of distance and time can be used, provided that they are consistent. But it is essential that the unit for angular speed is always radians per unit time.

If the speed of the particle round the circle is not constant, the definition of angular speed can be generalised as the rate of increase of the angle θ with respect to time, measured by the derivative $\dfrac{\mathrm{d}\theta}{\mathrm{d}t}$. But in this chapter all the examples are about objects moving with constant angular speed.

EXAMPLE 6.2.1

A car tachometer records the engine speed as 3000 revolutions per minute. What is this in $\mathrm{rad\,s^{-1}}$?

Each revolution is 2π radians, and the engine makes 50 revolutions in each second. The angular speed is therefore $50 \times 2\pi\,\mathrm{rad\,s^{-1}}$, which is $314\,\mathrm{rad\,s^{-1}}$, correct to 3 significant figures.

EXAMPLE 6.2.2

Taking the orbit of the earth round the sun to be a circle of radius 1.495×10^{11} metres, calculate the speed at which the earth is moving.

The earth completes the orbit in a year, which is approximately $365\frac{1}{4}$ days taking leap years into account. This is $365.25 \times 24 \times 60 \times 60$ seconds, or about 3.155×10^7 seconds. So the angular speed of the earth about the sun is $\dfrac{2\pi}{3.155 \times 10^7}$ rad s^{-1} $= 1.991\ldots \times 10^{-7}$ rad s^{-1}.

Using the equation $v = r\omega$, the speed of the earth is
$$\left(1.495 \times 10^{11}\right) \times \left(1.991\ldots \times 10^{-7}\right) \text{ms}^{-1} \approx 29\,766 \text{ ms}^{-1}.$$

The speed of the earth round the sun is just under $30\,000$ ms^{-1}.

EXAMPLE 6.2.3

The pilot of an aircraft flying at 800 km per hour on a bearing of 250° receives orders to change course to 210°. The manoeuvre is completed in 20 seconds. Calculate the radius of the turn.

The direction of flight changes by 40° in 20 seconds, or 2° in each second. The radius from the centre of the circle is always at right angles to the direction of flight, so this radius rotates at 2° per second, which is $\frac{1}{90}\pi$ rad s^{-1}.

The speed of the aircraft in km s^{-1} is $\dfrac{800}{60 \times 60}$, which is more simply written as $\frac{2}{9}$. So, using the equation $v = r\omega$,
$$r = \frac{v}{\omega} = \frac{2/9}{\pi/90} = \frac{20}{\pi} = 6.37, \text{ correct to 3 significant figures.}$$

Since the unit of v is km s^{-1}, the unit of r is km.

The radius of the turn is 6.37 km, correct to 3 significant figures.

Exercise 6A

1 A particle moves in a circle of radius 2 metres with speed 3 ms^{-1} Calculate its angular speed.

2 A cyclist completes a circuit of a circular track in 14 seconds. Calculate her angular speed in rad s^{-1}.

3 A model train moves round a circular track of diameter 1 metre in 3.7 seconds. Calculate the angular and linear speeds of the train.

4 A metre is approximately one ten-millionth of the distance from the north pole to the equator. Assuming that the earth is a sphere and that its axis of rotation is stationary, calculate the speed of a building on the equator in m s^{-1}.

5 The speed of an object moving round a circle is given as 8 m s^{-1} or 4 rad s^{-1}. Calculate the radius of its path.

6 An athlete runs round the semicircular end of the track in 12 seconds at a speed of 7 m s^{-1}. Calculate his angular speed, and the radius of the semicircle.

7   An aircraft changes its direction of motion from bearing 330° to bearing 120° in 4 seconds. Calculate two possible angular speeds for its turn.

8   An old gramophone record of diameter 30 cm rotates at 78 revolutions per minute. Calculate its angular speed in rad s^{-1}, and the speed of the rim of the record in m s^{-1}.

9   A train travelling at $180\,\text{km h}^{-1}$ moves round a circle of radius 600 metres. Calculate the angular speed of the train in rad s^{-1}.

10   A disc spins 15 times each second. Calculate the distance in centimetres from the axis of rotation of a point moving with speed $2\,\text{m s}^{-1}$.

6.3   Calculating the acceleration

Knowing r and ω, or r and v, completely determines the motion of a particle moving in a circle at constant speed. So it must be possible to find a formula for the acceleration in terms of either of these pairs of quantities. In this section the results will just be stated; a proof is given in the next section.

In terms of r and ω, the acceleration of the particle towards the centre of the circle is $r\omega^2$. Since ω can be written as $\dfrac{v}{r}$, this formula can also be written as $r \times \left(\dfrac{v}{r}\right)^2$, which is $\dfrac{v^2}{r}$. It is worth knowing the result in both forms, since it is sometimes more convenient to use angular speed and sometimes tangential speed.

> **A particle moving round a circular path of radius r with constant angular speed ω and tangential speed v has acceleration of magnitude $r\omega^2$ or $\dfrac{v^2}{r}$, directed towards the centre of the circle.**

The unit of acceleration given by these formulae depends on the units chosen for distance and time. If basic SI units are used for r and v, and if ω is in rad s^{-1}, the acceleration is in m s^{-2}.

EXAMPLE 6.3.1

Astronauts are trained to withstand the effects of high acceleration in a centrifugal machine. They sit or lie in cabins at the end of long metal arms, which rotate them about a vertical axis in a horizontal circle. If the radius of the circle is 12 metres, and the acceleration to be experienced is $10g$, how long should it take for the arms to make one revolution?

To produce an acceleration of $10g$, which is about $100\,\text{m s}^{-2}$ the angular speed ω must satisfy the equation $12\omega^2 = 100$, so $\omega = \sqrt{\dfrac{100}{12}}$.

The time in seconds for a complete rotation is $\dfrac{2\pi}{\omega}$, which is $2\pi\sqrt{\dfrac{12}{100}}$, or about 2.18.

The arm must make one revolution in about 2.2 seconds.

EXAMPLE 6.3.2

Compare the acceleration of the moon towards the centre of the earth with the acceleration due to gravity at the earth's surface.

The moon describes an approximately circular path of radius 3.844×10^8 m in 27.32 days, which is $27.32 \times 24 \times 60 \times 60$ s, or 2.36×10^6 s. The angular speed is therefore $\dfrac{2\pi}{2.36 \times 10^6}$ rad s^{-1} which is about 2.66×10^{-6} rad s^{-1}.

The acceleration of the moon towards the earth is therefore

$$\left(3.844 \times 10^8\right) \times \left(2.66 \times 10^{-6}\right)^2 \text{ ms}^{-2} \approx 0.002\,72 \text{ ms}^{-2}.$$

Comparing this with g, using the closer approximation $g \approx 9.8$ instead of $g \approx 10$, the acceleration of the moon is $0.000\,278g$, which is about $\frac{1}{3600}g$.

This calculation forms the basis of Newton's law of gravitation. The radius of the moon's orbit is about 60 times the radius of the earth, and (since $\sqrt{3600} = 60$) the acceleration of gravity at moon distance is about $\dfrac{1}{60^2}$ of the acceleration on the earth's surface. This suggests that the acceleration due to gravity is proportional to the inverse square of the distance from the earth.

EXAMPLE 6.3.3

A smooth circular table of radius 1.2 metres has a raised rim. A ball-bearing of mass 50 grams runs round the rim of the table, making one circuit every 4 seconds. Find the magnitude of the contact force on the ball-bearing from the rim.

The ball-bearing rotates through 2π radians about the centre of the table in 4 seconds, an angular speed of $\frac{1}{2}\pi$ rad s^{-1}. The acceleration of the ball-bearing towards the centre is therefore $1.2 \times \left(\frac{1}{2}\pi\right)^2$ ms^{-2}, which is $0.3\pi^2$ m s^{-2}.

The mass of the ball-bearing is 0.05 kg, so the force needed to produce the acceleration is $0.05 \times 0.3\pi^2$ N $= 0.148$ N, correct to 3 significant figures.

EXAMPLE 6.3.4

A racing cyclist travels round a circle of radius 30 metres at a speed of 15 m s^{-1}. What must be the coefficient of friction between the tyres and the track for this to be possible?

Fig. 6.3 shows the forces on the cyclist. The $\dfrac{v^2}{r}$ form of the acceleration formula shows that he requires an acceleration towards the centre of $\dfrac{15^2}{30}$ ms^{-2} which is 7.5 m s^{-2}.

The force which produces this acceleration is the friction between the tyres and the track. The mass of the cyclist with his machine is not given, so denote it by M kg. Then the friction force is $75M$ newtons.

The normal contact force from the ground is equal to the weight, which is $10M$ newtons.

So if the tyres are not to slip, the coefficient of friction must not be less than $\dfrac{7.5M}{10M}$, which is 0.75.

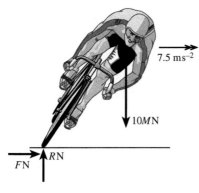

Fig. 6.3

6.4 Proof of the acceleration formula

You can if you wish omit this section and go straight to Exercise 6B.

When a particle moves along a straight line, the velocity is the derivative of the displacement with respect to time, and the acceleration is the derivative of the velocity. The equations for the motion of a projectile in Chapter 1 show that, if displacement, velocity and acceleration are regarded as vectors, the same relation holds for motion in two dimensions with constant acceleration. That is,

- the relation of the acceleration to the velocity is the same as the relation of the velocity to the displacement.

This principle is the basis from which the acceleration in circular motion can be found.

Fig. 6.4 shows a particle P moving round a circle centre O with angular speed ω. The displacement vector $\overrightarrow{OP}$ is denoted by $\mathbf{r}$, and the velocity vector $\mathbf{v}$ is perpendicular to $\mathbf{r}$.

Fig. 6.5 shows the relation between the velocity and acceleration vectors in a similar way. To do this, the velocity vector has its tail anchored at a point C, so that the displacement $\overrightarrow{CQ}$ represents the velocity. Since the velocity has constant magnitude, the point Q moves round a circle.

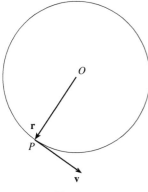

Fig. 6.4

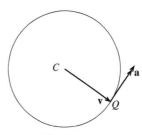

Fig. 6.5

Now the direction of $\overrightarrow{CQ}$ is always $\frac{1}{2}\pi$ radians in advance of the direction of $\overrightarrow{OP}$, so the angular speed of Q about C is also ω.

This has two consequences. First, since the direction of $\mathbf{v}$ is $\frac{1}{2}\pi$ radians in advance of $\mathbf{r}$, and the direction of $\mathbf{a}$ is $\frac{1}{2}\pi$ radians ahead of $\mathbf{v}$, it follows that $\mathbf{a}$ is π radians ahead of $\mathbf{r}$. That is, the direction of the acceleration of the particle is towards O.

Secondly, since the magnitude of $\mathbf{v}$ is $r\omega$, that is ω times the magnitude of $\mathbf{r}$, the magnitude of $\mathbf{a}$ is ω times the magnitude of $\mathbf{v}$, that is $(r\omega)\omega$, or $r\omega^2$. This completes the proof of the result stated in Section 6.3.

Exercise 6B

1 A child sits on a roundabout in a playground. The child is sitting 1.5 m from the centre of the roundabout, and the roundabout is turning at a constant rate of one revolution every 3 seconds.

   a Find the angular speed of the roundabout.

   b Find the speed of the child.

   c Find the acceleration of the child.

2 A particle P of mass 0.3 kg is attached to one end of a light inextensible string of length 0.6 metres. The other end of the string is attached to a fixed point O on a smooth horizontal surface. P moves in a circular path on the surface with speed 4 m s^{-1}. Calculate the tension in the string.

   Given that the tension in the string cannot exceed 30 N, find the maximum speed of P in m s^{-1}, and the angular speed in rad s^{-1}.

3 A particle P of mass 0.4 kg moves on a horizontal circle, centre O. The speed of the particle is 3 m s^{-1} and the force on P directed towards O is 15 N. Calculate the distance OP.

4 A toy train of mass 200 g is travelling on its tracks at the speed of 20 cm s^{-1}. It reaches a curved part of the track, which forms an arc of a horizontal circle of radius 60 cm. Calculate the horizontal force required to keep the train on its tracks as it travels along this curved part.

5 A radial force of 20 N is required to maintain a particle moving in a horizontal circle of diameter 1.8 metres with speed 4.8 m s^{-1}. Calculate the mass of the particle.

6 An object of mass 0.2 kg is placed on a horizontal turntable at a distance of 15 cm from the axis of rotation. When the turntable has angular speed 5 rad s^{-1}, friction between the object and turntable is limiting. Calculate the coefficient of friction.

7 A centrifuge consists of a hollow cylinder of diameter 0.8 metres rotating about a vertical axis with angular speed 500 rad s^{-1}. Calculate the magnitude of the contact force between the cylinder and an object of mass 0.7 kg on the inner surface of the centrifuge.

8 A light rod of length 1.2 metres is freely pivoted at one end to a fixed point O on a smooth horizontal surface. Particles P and Q, each of mass 0.3 kg, are attached to the mid-point and the free end of the rod respectively. The rod rotates in a horizontal circle at constant angular speed. Given that the tension in the rod between P and Q is 18 N, calculate the force on the rod at O.

9   A fairground ride consists of a horizontal wheel of radius 6 m in which participants stand against the wall of the wheel. During the first part of the ride, the wheel spins in a horizontal plane until the acceleration of the participants is equal to 0.8*g*. At this point, what is the speed of the wheel in revolutions per minute?

10 Two particles *P* and *Q*, of masses *M* and *m* respectively, are attached to opposite ends of a light rod of length *a*. The system is set rotating on a smooth horizontal surface, with the particles moving in concentric circles, centre *O*, about an axis perpendicular to the rod and the table. Find *OP*.

11 The acceleration due to gravity at a distance *r* metres from the centre of the earth is $\dfrac{k}{r^2}$ ms^{-2} where *k* is a constant. Estimate the height of a geostationary satellite with an orbit above the earth's equator, given that the radius of the earth is 6400 km. (A geostationary satellite is one which moves so that it is always above the same point on the earth's surface.)

12 The earth may be modelled as a uniform sphere of radius 6400 km, rotating on an axis passing through the north and south poles. Objects are released from points just above the earth's surface at the north pole and the equator. Estimate the difference in their acceleration towards the centre of the earth.

6.5   Three-dimensional problems

If an airliner has to change course, then the pilot banks the aircraft so that the wings are at an angle to the horizontal (see Fig. 6.6). The reason for this is to enable the lift force, which in level flight acts vertically and balances the weight, to act at an angle to the vertical. In this way the force can perform two functions at the same time: its vertical component balances the weight, and its horizontal component provides the required acceleration towards the centre of the circular path which the pilot wants to follow.

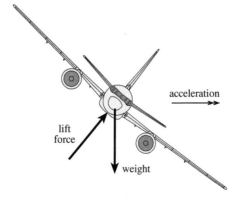

Fig. 6.6

You can use this principle in a number of applications, in which the supporting force may take the form of the tension in a string or the contact force from a surface. The examples in this section show how the principle of resolving can be applied to objects which move in a horizontal circle.

EXAMPLE 6.5.1

One end of a string of length l is tied to a hook, and a particle of mass m is attached to the other end. With the string taut and making an angle α with the downward vertical, the particle is set in motion so that it rotates in a horizontal circle about the vertical line through the hook. Find the period of one revolution of the particle round the circle.

There are just two forces on the particle, its weight mg and the tension T in the string (see Fig. 6.7). Suppose that the angular speed of the particle is ω.

The radius of the circle is $l\sin\alpha$, so the acceleration of the particle is $(l\sin\alpha)\omega^2$. You can now resolve horizontally and vertically to obtain equations connecting the various quantities.

$$\mathcal{R}(\rightarrow) \qquad T\cos\left(\tfrac{1}{2}\pi-\alpha\right)=m(l\sin\alpha)\omega^2,$$

$$\mathcal{R}(\uparrow) \qquad T\cos\alpha-mg=0.$$

Since $\cos\left(\tfrac{1}{2}\pi-\alpha\right)=\sin\alpha$, the first equation reduces to

$$T=ml\omega^2.$$

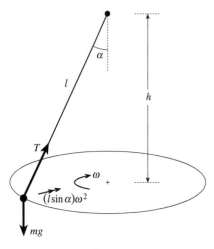

Fig. 6.7

Substituting this expression for T in the second equation gives

$$ml\omega^2\cos\alpha=mg,\text{ so } \omega=\sqrt{\frac{g}{l\cos\alpha}}.$$

The time for the particle to describe one revolution of 2π radians is $\dfrac{2\pi}{\omega}$, which is equal to $2\pi\sqrt{\dfrac{l\cos\alpha}{g}}$.

The apparatus in this example is called a **conical pendulum**, because the string describes the surface of a cone as it rotates.

It is interesting to note that $l\cos\alpha$ is just the depth h of the circular path below the hook, so that the period of one revolution can be written as $2\pi\sqrt{\dfrac{h}{g}}$. This expression is independent of l and of α, and also of m. This means that if you have a number of particles (not necessarily of the same mass) attached to strings of different lengths, all moving in circular paths at the same depth below the hook, they will all take the same time to make a complete revolution.

EXAMPLE 6.5.2

A bob-sleigh with its two-person team has a total mass of 200 kg. On one stretch of the course the team rounds a horizontal bend of radius 25 metres at a speed of 35 m s⁻¹. They bank the sleigh so that it rounds the bend with no sideways frictional force. Calculate the acceleration of the sleigh, and find the angle to the horizontal at which the sleigh is banked.

The acceleration towards the centre of the bend is $\dfrac{35^2}{25}\,\text{m s}^{-2}$ which is 49 m s⁻².

Fig. 6.8 shows the forces on the sleigh as it rounds the bend. Since there is no sideways frictional force, the acceleration towards the centre of the bend is provided entirely by the resolved part of the normal contact force of R newtons.

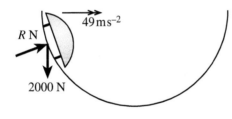

Fig. 6.8

If the angle at which the sleigh is banked is $\alpha°$,

$\mathcal{R}(\uparrow)\qquad R\cos\alpha° - 2000 = 0,$ and

$\mathcal{R}(\rightarrow)\qquad R\sin\alpha° = 200 \times 49.$

This gives

$$\tan\alpha° = \frac{R\sin\alpha°}{R\cos\alpha°} = \frac{200\times49}{2000} = 4.9, \text{so } \alpha \approx 78.5.$$

The sleigh is banked at an angle of about 78° to the horizontal.

Exercise 6C

1 A light inextensible string of length 1.2 m is attached to a fixed point on a ceiling. A particle of mass 0.8 kg is attached to the end of the string. It moves at constant speed in a horizontal circle of radius 0.4 m with centre vertically below the fixed point. Calculate, in either order,

  a the speed of the particle,

  b the tension in the string.

2 A light inextensible string of length 0.9 m is attached to a fixed point A. A particle of mass 1.3 kg is attached to the end of the string. It moves in a horizontal circle at constant speed with centre O vertically below A. The tension in the string is 14.6 N. Find

   a   the angle between the string and the vertical,

   b   the angular speed of the particle.

3 One end of a light inextensible string of length 0.6 metres is attached to a fixed point A, 0.3 metres above a smooth horizontal surface. The other end of the string is attached to a particle P of mass 0.4 kg. The particle moves with constant speed in a horizontal circle, with the string taut, and making an angle $\theta°$ with the vertical.

   a   Show that, when $\theta = 70$, the particle is not in contact with the surface, and calculate the tension in the string and the angular speed of P.

   b   When $\theta = 60$, the contact force between the particle and the surface is zero, and the angular speed of P is ω rad s^{-1}. Calculate the tension in the string and the value of ω.

   c   With $\theta = 60$, P moves with angular speed 2 rad s^{-1}. Calculate the tension in the string, and the contact force between the particle and the surface.

4

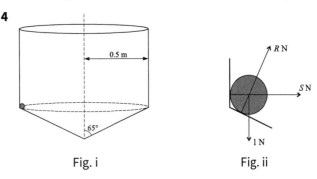

Fig. i   Fig. ii

A hollow container consists of a smooth circular cylinder of radius 0.5 m, and a smooth hollow cone of semi-vertical angle 65° and radius 0.5 m. The container is fixed with its axis vertical and with the cone below the cylinder. A steel ball of weight 1 N moves with constant speed 2.5 ms^{-1} in a horizontal circle inside the container. The ball is in contact with both the cylinder and the cone (see Fig. i). Fig. ii shows the forces acting on the ball, i.e. its weight and the forces of magnitudes R N and S N exerted by the container at the points of contact. Given that the radius of the ball is negligible compared with the radius of the cylinder, find R and S.

   (Cambridge International AS and A Level Mathematics 9709/05 Paper 5 Q3
   June 2007)

5 A particle P is moving inside a smooth hollow cone which has its vertex downwards and its axis vertical, and whose semi-vertical angle is 45°. A light inextensible string parallel to the surface of the cone connects P to the vertex. P moves with constant angular speed in a horizontal circle of radius 0.67 m (see diagram). The tension in the string is equal to the weight of P. Calculate the angular speed of P.

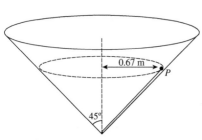

   (Cambridge International AS and A Level Mathematics 9709/51 Paper 5
   Q4 November 2012)

6   A particle P of mass 0.2 kg is attached to a fixed point A by a light inextensible string of length 0.4 m. A second light inextensible string of length 0.3 m connects P to a fixed point B which is vertically below A. The particle P moves in a horizontal circle, which has its centre on the line AB, with the angle $APB = 90°$ (see diagram).

   i   Given that the tensions in the two strings are equal, calculate the speed of P.

   ii It is given instead that P moves with its least possible angular speed for motion in this circle. Find this angular speed.

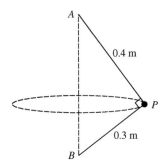

(Cambridge International AS and A Level Mathematics 9709/51 Paper 5
Q5 November 2013)

7   A narrow groove is cut along a diameter in the surface of a horizontal disc with centre O. Particles P and Q, of masses 0.2 kg and 0.3 kg respectively, lie in the groove, and the coefficient of friction between each of the particles and the groove is μ. The particles are attached to opposite ends of a light inextensible string of length 1 m. The disc rotates with angular velocity ω rad s^{-1} about a vertical axis passing through O and the particles move in horizontal circles (see diagram).

   i   Given that $\mu = 0.36$ and that both P and Q move in the same horizontal circle of radius 0.5 m, calculate the greatest possible value of ω and the corresponding tension in the string.

   ii Given instead that $\mu = 0$ and that the tension in the string is 0.48 N, calculate

       a   the radius of the circle in which P moves and the radius of the circle in which Q moves,

       b   the speeds of the particles.

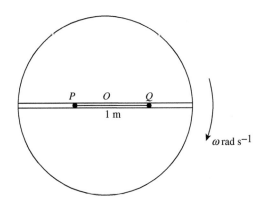

(Cambridge International AS and A Level Mathematics 9709/51 Paper 5
Q7 June 2011)

8 An aircraft of mass 2 tonnes flies at $540 \, \text{km} \, \text{h}^{-1}$ in a horizontal circular arc to change its direction of motion from bearing 320° to bearing 046°. This manoeuvre is executed in 30 seconds, with the aircraft banked at $\alpha°$ to the horizontal. Calculate two values of α and the corresponding values of the lift force perpendicular to the surface of the aircraft's wings.

Miscellaneous exercise 6

1 A horizontal circular disc rotates with constant angular speed $9 \, \text{rad} \, \text{s}^{-1}$ about its centre O. A particle of mass $0.05 \, \text{kg}$ is placed on the disc at a distance $0.4 \, \text{m}$ from O. The particle moves with the disc and no sliding takes place. Calculate the magnitude of the resultant force exerted on the particle by the disc.

> (Cambridge International AS and A Level Mathematics 9709/51 Paper 5
> Q1 November 2010)

2 The end A of a rod AB of length $1.2 \, \text{m}$ is freely pivoted at a fixed point. The rod rotates about A in a vertical plane. Calculate the angular speed of the rod at an instant when B has speed $0.6 \, \text{m} \, \text{s}^{-1}$.

> (Cambridge International AS and A Level Mathematics 9709/51 Paper 5
> Q1 June 2012)

3 One end of a light inextensible string of length $0.16 \, \text{m}$ is attached to a fixed point A which is above a smooth horizontal table. A particle P of mass $0.4 \, \text{kg}$ is attached to the other end of the string. P moves on the table in a horizontal circle, with the string taut and making an angle of 30° with the downward vertical through A (see diagram). P moves with constant speed $0.6 \, \text{m} \, \text{s}^{-1}$. Find

i the tension in the string,

ii the force exerted by the table on P.

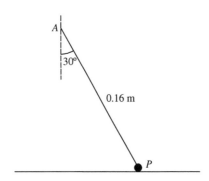

> (Cambridge International AS and A Level Mathematics 9709/05 Paper 5
> Q2 November 2007)

4   A hollow cylinder of radius 0.35 m has a smooth inner surface. The cylinder is
    fixed with its axis vertical. One end of a light inextensible string of length 1.25 m
    is attached to a fixed point O on the axis of the cylinder. A particle P of mass
    0.24 kg is attached to the other end of the string. P moves with constant speed in
    a horizontal circle, in contact with the inner surface of the cylinder, and with the
    string taut (see diagram).

    i   Find the tension in the string.

    ii Given that the magnitude of the acceleration of P is 8 m s^{-2}, find the force
            exerted on P by the cylinder.

    (Cambridge International AS and A Level Mathematics 9709/05 Paper 5
    Q3 November 2006)

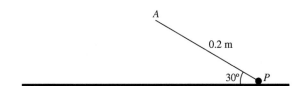

5   One end of a light inextensible string of length 0.2 m is attached to a fixed point A
    which is above a smooth horizontal surface. A particle P of mass 0.6 kg is attached
    to the other end of the string. P moves in a circle on the surface with constant
    speed v m s^{-1}, with the string taut and making an angle of 30° to the horizontal
    (see diagram).

    i   Given that $v = 1.5$, calculate the magnitude of the force that the surface exerts on P.

    ii Given instead that P moves with its greatest possible speed while remaining in
            contact with the surface, find v.

A

0.2 m

30° •P

    (Cambridge International AS and A Level Mathematics 9709/51 Paper 5
    Q3 November 2010)

6   A particle of mass 0.12 kg is moving on the smooth inside surface of a fixed hollow
    sphere of radius 0.5 m. The particle moves in a horizontal circle whose centre is
    0.3 m below the centre of the sphere (see diagram).

    i   Show that the force exerted by the sphere on the particle has magnitude 2 N.

    ii   Find the speed of the particle.

    iii Find the time taken for the particle to complete one revolution.

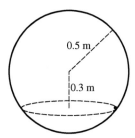

0.5 m

0.3 m

    (Cambridge International AS and A Level Mathematics 9709/05 Paper 5
    Q4 June 2009)

121

7   One end of a light inextensible string of length 0.7 m is attached to a fixed point A. The other end of the string is attached to a particle P of mass 0.25 kg. The particle P moves in a circle on a smooth horizontal table with constant speed 1.5 m s^{-1}. The string is taut and makes an angle of 40° with the vertical (see diagram). Find

i   the tension in the string,

ii   the force exerted on P by the table.

P now moves in the same horizontal circle with constant angular speed ω rad s^{-1}.

iii   Find the maximum value of ω for which P remains on the table.

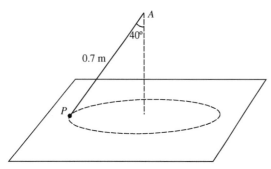

(Cambridge International AS and A Level Mathematics 9709/51 Paper 5 Q6 November 2009)

8   One end of a light inextensible string of length 0.2 m is attached to a fixed point A which is above a smooth horizontal table. A particle P of mass 0.3 kg is attached to the other end of the string. P moves on the table in a horizontal circle, with the string taut and making an angle of 60° with the downward vertical (see diagram).

i   Calculate the tension in the string if the speed of P is 1.2 m s^{-1}.

ii   For the motion as described, show that the angular speed of P cannot exceed 10 rad s^{-1}, and hence find the greatest possible value for the kinetic energy of P.

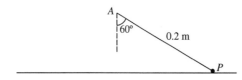

(Cambridge International AS and A Level Mathematics 9709/51 Paper 5 Q6 June 2013)

9   A smooth bead B of mass 0.3 kg is threaded on a light inextensible string of length 0.9 m. One end of the string is attached to a fixed point A, and the other end of the string is attached to a fixed point C which is vertically below A. The tension in the string is T N, and the bead rotates with angular speed ω rad s^{-1} in a horizontal circle about the vertical axis through A and C.

i   Given that B moves in a circle with centre C and radius 0.2 m, calculate ω, and hence find the kinetic energy of B.

ii   Given instead that angle $ABC = 90°$, and that AB makes an angle $\tan^{-1}\left(\dfrac{1}{2}\right)$ with the vertical, calculate T and ω.

(Cambridge International AS and A Level Mathematics 9709/51 Paper 5 Q6 November 2011)

10 A bend in a horizontal road has a radius of 200 metres. To improve safety, it is decided to bank the road surface at an angle $\alpha°$ to the horizontal. It is required that a vehicle can remain stationary on the road when the surface is icy, and the coefficient of friction is 0.052. Calculate the greatest permissible value of α, and the corresponding greatest safe speed, in $km\,h^{-1}$, of a vehicle when the road surface is icy.

Chapter 7
Geometrical methods

Conditions for the equilibrium of rigid objects can also be expressed in geometrical form.
When you have completed this chapter, you should

- know the parallelogram rule for combining forces on a rigid object
- know and be able to apply the conditions for equilibrium of a rigid object acted on by three non-parallel forces
- understand how the standard model of friction can be recast in terms of a total contact force and the angle of friction
- be able to use geometrical methods to solve problems involving friction, including those in which equilibrium may be broken by either sliding or toppling.

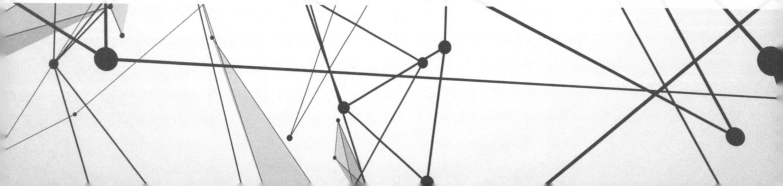

7.1 The parallelogram rule

A narrow boat 20 metres long is being manoeuvred into a mooring by a river bank. Ropes are attached to the boat at the bow and the stern, and these are held by two children on the bank. The girl with the bow rope exerts a force of 40 newtons at an angle 50° to the direction in which the boat is pointing; the boy with the stern rope exerts a force of 30 newtons at right angles to the boat. (See Fig. 7.1.) Suppose that only one person was available. Where would the rope have to be attached, and what force should be exerted, to produce the same effect on the boat as the two children?

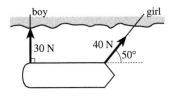

Fig. 7.1

If the boat could be modelled as a particle, you would use the triangle rule to combine the two forces, as in Fig. 7.2. You can calculate that this gives a force of 65.9 newtons at 67° to the direction of the boat.

Fig. 7.2

But this boat can't adequately be modelled as a particle. Where the rope is attached makes a lot of difference to how the boat behaves. So the triangle law by itself is not enough. In Fig. 7.3 the lines of the two ropes are produced backwards to meet at a point X. Since the lines of action of both forces pass through X, the forces have no moment about X. It follows that a single force with the same effect must also have no moment about X, so that its line of action has to pass through X.

So if you draw a line through X at an angle of 67°, a force of 65.9 newtons along this line has the same effect as the original two forces. If this line cuts the side of the boat at Y, as in Fig. 7.3, then the single rope should be attached at Y. You can calculate that Y is 10.1 metres from the stern of the boat.

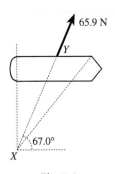

Fig. 7.3

The two parts of this calculation can be amalgamated by replacing the triangle law for combining forces by a parallelogram law, as in Fig. 7.4. Instead of drawing a triangle with the tail of the second arrow at the head of the first, you can draw a parallelogram with the tails of both arrows at X. An arrow along the diagonal with its tail at X then represents the resultant force on the boat in magnitude, direction and line of action.

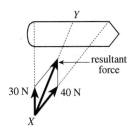

Fig. 7.4

> The parallelogram rule for combining forces **If two forces P and Q, acting along lines which intersect at X, are represented by arrows on some scale with their tails at X which define a parallelogram, and the arrow representing R is the diagonal with its tail at X, then the single force R has exactly the same effect on a rigid object as the two forces P and Q acting together.**

Exercise 7A

1 AB is a girder 10 metres long. Cables are attached to it at A and B, making angles of $\alpha°$ and $\beta°$ with the girder, and tension forces of PN and QN are applied, as shown in the figure. The effect on the girder is the same as that of a single force of RN at an angle $\theta°$ to the girder applied at a point X. By drawing and measurement, find the values of R and θ and the distance AX in the following cases.

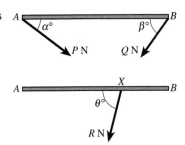

   a $P = 700$, $Q = 1000$, $\alpha = 90$, $\beta = 60$

   b $P = 1000$, $Q = 1200$, $\alpha = 45$, $\beta = 65$

   c $P = 800$, $Q = 1400$, $\alpha = 140$, $\beta = 160$

125

2 With the notation of Question 1, use a parallelogram of forces to calculate the values of R and θ and the distance AX in the following cases.

   a $P = 1000$, $Q = 500$, $\alpha = 180$, $\beta = 45$

   b $P = 1000$, $Q = 1000$, $\alpha = 130$, $\beta = 30$

   c $P = 1500$, $Q = 1000$, $\alpha = 155$, $\beta = 115$

   d $P = 1000$, $Q = 2000$, $\alpha = 80$, $\beta = 50$

3 $ABCD$ is a rectangular table with $AB = 3$ metres and $BC = 2$ metres. Two people push the table with forces of equal magnitude. One pushes at A in the direction of the edge AB, and the other pushes at a point X of DC, in such a way that the resultant force acts through the centre of the table at an angle of $30°$ to AB. Find the distance DX.

4 Two boys stand at the ends of a diameter AB of a circular table. They both push the table with forces of 50 newtons along the tangents at A and B to try to spin the table clockwise. A third boy, standing at a point C halfway round the circumference from A to B, tries to stop the spin by pushing with a force of 100 newtons anticlockwise. By finding the resultant of two of the forces, and then combining this with the third force, find the magnitude and the line of action of the combined effect of all three forces.

7.2   Three forces in equilibrium

Similar arguments can be used when there are just three forces acting on a rigid object in equilibrium. If these forces are not all parallel, let the lines of action of two of them meet at X. Then the moments of these two forces about X are 0. Since the moments of all three forces about X have to balance, the moment of the third force about X must also be 0. This means that the line of action of the third force also goes through X.

Now the conditions for the object to be in equilibrium can be expressed by resolving in two directions and taking moments about one point. You already know that, for a particle acted on by three forces, the resolving equations are equivalent to the triangle of forces rule. (See M1 Section 10.5.) This applies equally well for the forces on a rigid object.

These two ideas can be put together to give conditions for three forces on a rigid object to be in equilibrium, provided that they are not parallel.

> **A rigid object acted on by just three non-parallel forces is in equilibrium if**
>
> - **the forces can be represented in a triangle of forces, and**
>
> - **the lines of action of the forces are concurrent.**

You can use this instead of the algebraic methods described in Chapter 4. It is not so general, because it only applies when there are just three forces and these are not parallel. But in this case it often produces more efficient solutions.

EXAMPLE 7.2.1

A boy is designing a go-kart and wants to know where its centre of mass is. He finds that his kart will rest horizontally with the brakes off if the rear wheels are on a 15° slope and the front wheels on a 22° slope. What does this tell him about the centre of mass?

Fig. 7.5 shows that there are only three forces on the kart, its weight and the contact forces on the front and rear wheels. The lines of action must be concurrent, so the centre of mass G lies on a vertical line through the point where the lines of the two contact forces meet.

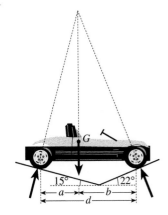

Fig. 7.5

Suppose that this line cuts the line joining the axles at a distance a in front of the rear wheels and b behind the front wheels. Then the contact forces intersect at a point whose height above the axles can be expressed either as $a \tan 75°$ or $b \tan 68°$. Therefore $a \tan 75° = b \tan 68°$, so $\dfrac{a}{b} = \dfrac{\tan 68°}{\tan 75°} \approx 0.66$, or $a \approx \frac{2}{3} b$

So if the distance between the axles is d, the centre of mass is on a vertical line about $\frac{2}{5} d$ in front of the rear axle.

Notice that in this example you don't want to find the contact forces, so the concurrency condition tells you all you need to know.

EXAMPLE 7.2.2

A window of weight 200 newtons is hinged along its top edge. Its centre of mass is at its geometrical centre. It is kept open at 40° to the vertical by the thrust from a light strut perpendicular to the window and attached to the wall. Calculate the thrust in the strut and the force exerted on the window by the hinge.

When you use algebraic methods it is often best to represent a hinge force by its components in two perpendicular directions. But to use the geometrical conditions you want to reduce the number of forces to three. Two of these are the weight and the thrust in the strut, so you must treat the hinge force as a single force acting at an angle.

In Fig. 7.6 H is the hinge, G the centre of mass of the window, ML the strut and N the point where the vertical through G meets ML, which is the mid-point of ML. The thrust in the strut is T newtons, and the force from the hinge is R newtons. The length HL is not given, so denote it by d.

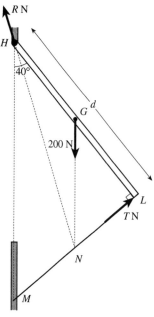

Fig. 7.6

By the concurrency condition, the line of action of the force $\mathbf{R}$ passes through N. You can easily calculate that $ML = d\tan 40°$, so $NL = \frac{1}{2}d\tan 40°$ and angle $NHL = \tan^{-1}(\frac{1}{2}\tan 40°)$, which is $22.7\ldots°$. Therefore $\mathbf{R}$ makes an angle of $40° - 22.7\ldots° = 17.2\ldots°$ with the vertical.

You now know all the angles in the triangle of forces, which is drawn in Fig. 7.7. By the sine rule,

$$\frac{T}{\sin 17.2\ldots°} = \frac{R}{\sin 50°} = \frac{200}{\sin 112.7\ldots°},$$

so $T = 64.3$ and $R = 166.1$, to 1 decimal place.

The thrust in the strut is about $64\,\text{N}$ and the force from the hinge is about $166\,\text{N}$ at $17°$ to the vertical.

This question could also be solved using Lami's theorem; the argument is very similar.

Fig. 7.7

In this example it is interesting to notice that in Fig. 7.6 the triangle HMN has sides parallel to the three forces, so this triangle is similar to the triangle of forces. Another way of completing the solution would then be to find the lengths MN, NH and HM, and to observe that T, R and 200 are proportional to these lengths. This does not make the calculation any easier, but it would be a good method to use if you were solving the problem by scale drawing.

Exercise 7B

Use the geometrical method described in Section 7.2 to work the problems in this exercise.

1 A uniform rod XY, of length 1 metre and mass 4 kg, is smoothly hinged at X to a vertical wall. The rod is kept in equilibrium, making an angle of 30° with the upward vertical, by a string attached to the rod at Y and to a hook in the wall at the same level as Y. Find the force on the rod from the hinge at X, and the tension in the string.

2 A loaded shelf AB is hinged to a wall at A. It is kept in a horizontal position by a light strut CD which supports it at a point D, where $AD = 12$ cm. The other end C of the strut is pinned to the wall, 16 cm below A. The total weight of the shelf and its load is 200 newtons, and its centre of mass is 30 cm horizontally from A. Find the direction and the magnitude of the force on the hinge at A.

3 Two panes of glass are laid on a table, touching each other along one edge. The opposite edges are then raised so that the panes are fixed at angles of 30° and 15° to the horizontal. A non-uniform beam AB, of length 2.5 metres, rests horizontally with one end on each pane and at right angles to the common edge. The end A is in contact with the steeper pane. Making the assumption that the contacts are smooth, find the distance of the centre of mass of the beam from A.

4 A uniform rod PQ has mass 3 kg and length 60 cm. The end P is pivoted at a fixed point and the end Q is attached to one end of a string. The other end of the string is attached to a fixed point R, which is in the same vertical plane as PQ. Angle PQR is a right angle, and PQ is inclined at 15° to the horizontal. Find the force exerted on the rod by the pivot at P

   a if Q is higher than P,   b if P is higher than Q.

5 A large uniform flag is suspended from a horizontal rail AB of length 6 m. The total weight of 500 newtons hangs by chains attached to hooks P and Q in the roof of a hall and to points X and Y of the rail, where $AX = 2$ m and $YB = 1$ m. Hook P is 4 m directly above A. Find the distance between the hooks, and the tension in the chains.

6 A uniform rod of weight 80 newtons is suspended from the ceiling by strings attached to its ends. The rod is in equilibrium at an angle of 10° to the horizontal, and the string attached to the higher end is at an angle of 40° to the vertical. Find the angle which the other string makes with the vertical. Find also the tension in the two strings.

7 Two smooth rails are set up in a vertical plane making angles of 20° and 70° with the horizontal. A uniform rod, with a ring at each end, is supported in equilibrium with one ring on each rail. Find the angle which the rod makes with the horizontal.

8 In Question 7, suppose that the rails are set up at 20° and 40° to the horizontal. Use the sine rule twice to show that the angle $\theta°$ made by the rod with the horizontal then satisfies the equation $\dfrac{\sin(70-\theta)°}{\sin 20°} = \dfrac{\sin(50+\theta)°}{\sin 40°}$, and hence find θ.

7.3 The angle of friction

The final two sections of this chapter describe how geometrical methods can be used in situations where there are frictional forces. You may if you wish omit these sections, since you already know one way of dealing with problems of this type. However, the geometric approach often produces neater solutions and gives new insights into the mechanical principles involved. In this section the method is applied to objects modelled as particles.

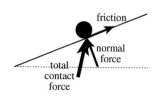

Fig. 7.8

Whenever you have a frictional force, there is always a normal contact force at right angles to it. The normal force and the friction can therefore be combined to give a resultant called the **total contact force**, as shown in Fig. 7.8.

Fig. 7.9 is the force diagram. The friction and the normal force have magnitudes F and R, and θ is the angle between the total contact force and the normal, so that

$$\tan\theta = \frac{F}{R}.$$

Fig. 7.9

This fraction is familiar! One of the properties of the frictional force is that $\dfrac{F}{R} \le \mu$, the coefficient of friction. You can now write this property in the form $\tan\theta \le \mu$, or $\theta \le \tan^{-1}\mu$.

The angle $\tan^{-1}\mu$ is called the **angle of friction**. It is usually denoted by the Greek letter λ (lambda). The friction property can then be restated in terms of the angle λ instead of the coefficient μ.

> **When one surface tends to slide over another, the angle between the total contact force and the normal is less than or equal to the angle of friction λ. When friction is limiting, the angle is equal to λ.**
>
> **The value of λ is $\tan^{-1}\mu$, a constant which depends on the nature and materials of the surfaces.**

Replacing the normal reaction and the frictional force by the total contact force reduces the number of forces by one. This often makes it possible to use the condition for the equilibrium of two or three forces, as described in M1 Section 10.5.

EXAMPLE 7.3.1 (SEE M1 EXPERIMENT 5.3.2)

Place a book on a table. Slowly tilt the table until the book starts to slide, and measure the angle of inclination of the table to the horizontal when this happens. If this angle is θ, show that the coefficient of friction is equal to $\tan\theta$.

Fig. 7.10 shows the only two forces acting on the book, its weight and the total contact force from the table. If the table is tilted at an angle θ, the weight is at an angle θ to the normal. Since the contact force acts in the direction opposite to the weight, it too makes an angle θ to the normal.

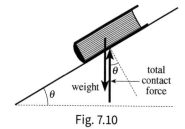

Fig. 7.10

In limiting equilibrium $\theta = \lambda$, so $\tan\theta = \tan\lambda = \mu$.

EXAMPLE 7.3.2 (SEE M1 EXAMPLE 5.5.1)

A block of weight 20 N is at rest on a horizontal surface. When a force of magnitude 12 N is applied to the block at an angle of 30° above the horizontal, it is on the point of moving. Find the coefficient of friction between the block and the surface.

In Fig. 7.11 the friction and normal reaction are combined as a total contact force. This reduces the number of forces on the object to three, so equilibrium can be expressed by a triangle of forces (Fig. 7.12). Since the object is on the point of moving, friction is limiting, so the contact force is at an angle $\lambda°$ to the vertical.

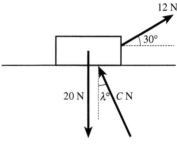

Fig. 7.11

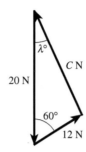

Fig. 7.12

Denote the magnitude of the contact force by CN. By the cosine rule,

$$C^2 = 20^2 + 12^2 - 2 \times 20 \times 12 \times \cos 60° = 400 + 144 - 240 = 304.$$

The sine rule then gives

$$\frac{C}{\sin 60°} = \frac{12}{\sin \lambda°}, \quad \text{so} \quad \sin \lambda° = \frac{12 \sin 60°}{\sqrt{304}}.$$

This gives $\lambda = 36.5...$, so that $\mu = \tan \lambda° = 0.74$, to 2 significant figures.

The coefficient of friction between the object and the surface is about 0.74.

EXAMPLE 7.3.3

A rough path climbs at an angle a to the horizontal. A gardener drags a heavy sack of weight W up the path at constant speed with a rope inclined at an angle β to the path, as shown in Fig. 7.13. For what value of β will the tension in the rope be smallest?

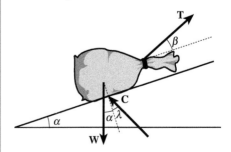

Fig. 7.13

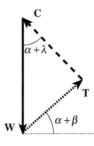

Fig. 7.14

Since the sack is moving, friction is limiting. The contact force C is at an angle λ to the normal, so it is at an angle $\alpha + \lambda$ to the vertical. At a steady speed the forces are in equilibrium, so a triangle of forces can be drawn, as in Fig. 7.14.

In this triangle the side representing **W** is completely known. You also know the direction of the side representing **C**, but its length depends on the choice of β. The aim is to make the length of the side representing the tension **T** as small as possible.

This will happen when the angle between **T** and **C** is a right angle. So to minimise the tension in the rope, you should make $\beta = \lambda$. That is, the rope should be at an angle λ to the path.

Exercise 7C

You should use a geometrical method to work the problems in this exercise.

1 An object of weight $50\,\text{N}$ is at rest on a horizontal plane. An upward force of magnitude $20\,\text{N}$ is applied to the object in a direction making an angle of $30°$ with the horizontal, as shown in the diagram. Given that the object is on the point of sliding, find

   a the magnitude of the total contact force between the object and the plane,

   b the angle of friction,

   c the coefficient of friction.

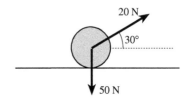

2 A crate of mass $100\,\text{kg}$ is to be moved from rest along horizontal ground. The angle of friction for the contact between the crate and the ground is $35°$. Find the magnitude of the force necessary to just move the crate when this force is

   a a push applied at $25°$ downwards from the horizontal,

   b a pull applied at $25°$ upwards from the horizontal.

3 A log of weight W rests on horizontal ground. When it is pulled with a force of magnitude F at an angle of $a°$ above the horizontal, the log is on the point of moving. The angle of friction for the contact between the log and the ground is $\lambda°$.

   a Show that $\dfrac{F}{W} = \dfrac{\sin\gamma°}{\cos(\gamma - \alpha)°}$.

   b Find the ratio $\dfrac{F}{W}$ when $\gamma = 25$ and $\alpha = 20$.

   c Show that $\dfrac{F}{W}$ is least when $\alpha = \gamma$.

4 A parcel of mass 6 kg rests on a plane inclined at 40° to the horizontal. The parcel is held in limiting equilibrium by a force of magnitude 15 N acting at 20° upwards from the plane, as shown in the diagram. Find

a the magnitude of the total contact force,

b the direction of the total contact force,

c the coefficient of friction between the parcel and the plane.

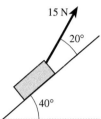

5 An adjustable inclined plane can be set at any angle to the horizontal between 0° and 90°. With the angle of inclination set at 15°, a block of weight 12 N is pulled up the incline at constant speed by a force parallel to the plane of magnitude 7 N. Find the direction of the total contact force acting on the block. Hence find the angle of inclination of the plane at which the block can rest, in the absence of any applied force, whilst on the point of sliding down the plane.

6 A ring of mass 2 kg is threaded on a fixed horizontal rod. It is pulled by a force of magnitude P N at an angle to the rod. The ring is on the point of moving along the rod. The coefficient of friction between the ring and the rod is 0.7. Find

a the angle the total contact force makes with the upward vertical,

b the value of P when the pulling force acts at an angle of 20° above the horizontal,

c the value of P when the pulling force acts at an angle of 20° below the horizontal.

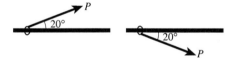

7 The diagram shows a man of weight 800 N using a power-saw to make a cut into a tree-trunk at an angle of 20° downwards from the horizontal. The magnitude of the force exerted on the man and his saw by the tree-trunk, in the vertical plane perpendicular to the cut, is denoted by P N. The contact force exerted on the man by the horizontal ground is denoted by **C**.

Find the magnitude and direction of **C** when

a $P = 200$, **b** $P = 300$.

In each case, state the magnitude and direction of the force exerted by the man and his saw on

i the tree-trunk, **ii** the horizontal ground.

Give a reason why the magnitude of the force exerted by the man and his saw on the tree-trunk cannot exceed $\dfrac{800}{\cos 70°}$ N.

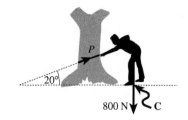

133

7.4 Rigid body problems involving friction

The method in Section 7.3 is even more powerful when the forces act on a rigid object, since the concurrency rule can often be used to find the angle which the total contact force makes with the normal. This angle cannot be greater than λ, the angle of friction.

EXAMPLE 7.4.1

A uniform rod of length l has a ring at one end which can slide along a rough horizontal rail. The coefficient of friction between the ring and the rail is μ. The other end of the rod is attached to the end of the rail by a cord, which is also of length l. What is the smallest angle which the rod can make with the horizontal?

Fig. 7.15 shows the three forces on the rod: its weight, the tension in the cord and the total contact force between the ring and the rail. Let the rod make an angle θ with the horizontal, and let μ be the angle which the contact force makes with the vertical. Since the rod and the cord have the same length, the cord also makes an angle θ with the horizontal.

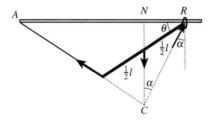

Fig. 7.15

The lines of action of the forces must be concurrent, at the point labelled C. In Fig. 7.15, R is the ring, and the line of action of the weight meets the rail at N. Then $RN = \frac{1}{2} l \cos\theta$. By symmetry, angle $CAN = \theta$ and $AC = l + \frac{1}{2} l = \frac{3}{2} l$, so $CN = \frac{3}{2} l \sin\theta$.

Therefore $\tan\alpha = \dfrac{RN}{CN} = \dfrac{1}{3}\cot\theta$.

If the ring is not to slide, $\tan\alpha$ must be less than or equal to μ, so that $\frac{1}{3}\cot\theta \le \mu$. Therefore $\theta \ge \cot^{-1} 3\mu$.

The smallest angle which the rod can make with the horizontal is $\cot^{-1} 3\mu$.

You can often use the method to find whether, as an applied force is increased, equilibrium is broken by sliding or toppling. For example, Fig. 7.16 shows again the box being pushed across the floor, which was discussed in Section 4.2, but now the separate normal and friction forces have been replaced by a total contact force.

There are three forces, so you can draw a triangle of forces as in Fig. 7.17. You can see from this diagram that, as the push is gradually increased, the angle α between the contact force and the vertical gets bigger.

You also know from Fig. 7.16 that, so long as the box is in equilibrium, the line of action of the contact force has to go through C, the middle of the top of the box. So, as α gets bigger, one of two things can happen. Either α reaches the value λ, in which case the box starts to slide; or the point of the base at which the contact force acts reaches the bottom right corner of the box, so that the box starts to topple.

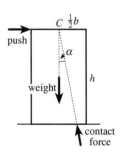

Fig. 7.16

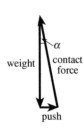

Fig. 7.17

The condition for sliding is therefore that $\tan\alpha = \mu$, and the condition for toppling is that $\tan\alpha = \dfrac{b}{2h}$. So, if you push hard enough, equilibrium is broken by sliding if $\mu < \dfrac{b}{2h}$, and by toppling if $\mu > \dfrac{b}{2h}$.

EXAMPLE 7.4.2

Use a geometrical method to solve Example 4.2.2.

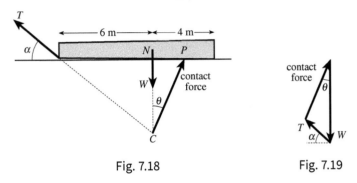

Fig. 7.18   Fig. 7.19

Fig. 7.18 shows the three forces with the normal force and the friction replaced by a total contact force at an angle θ to the vertical. From the triangle of forces (Fig. 7.19), since α remains constant, θ increases as T is increased.

If C is the point where the lines of action concur, and N the point of the tree trunk on the ground below the centre of mass, then $CN = 6\tan\alpha$ metres. So if the contact force acts at P, $NP = 6\tan\alpha\tan\theta$ so long as the trunk is in equilibrium.

Equilibrium can be broken by sliding if $\tan\theta$ reaches the value 0.6, and by toppling if P reaches the end of the trunk, that is if $6\tan\alpha\tan\theta = 4$. So the narrow end will be lifted off the ground if $6 \times 0.6\tan\alpha > 4$, that is if
$$\alpha > \tan^{-1}\frac{4}{3.6} \text{ or } \alpha > 48.0°, \text{ to 3 significant figures.}$$
To lift the trunk clear of the ground before it starts to drag, the cable should make an angle of at least 48° with the horizontal.

You can see from an example like this that the geometrical approach may not only lead to simpler mathematics, but also gives more insight into what happens as the force is increased. So although the limitation to three forces means that you can't always use geometrical methods, they are worth trying when you have a problem to which they can be applied.

Exercise 7D

You should use a geometrical method to work the problems in this exercise.

1 A uniform ladder has one end on rough horizontal ground and leans at the other end against a vertical window. It rests in equilibrium at 60° to the horizontal. The contact with the window is so smooth that no reliance can be placed on the friction force there. Find the smallest acceptable coefficient of friction between the ladder and the ground.

A painter, whose mass is four times that of the ladder, now climbs up to the window. Find the smallest acceptable coefficient of friction if the ladder is still not to slip.

2 A uniform ladder is placed on rough horizontal ground, and rests over the top of a vertical wall in such a way that one-quarter of its length projects above the wall. The contact between the ladder and the top of the wall is assumed to be smooth. The ladder rests in equilibrium at 60° to the horizontal. Find the least acceptable coefficient of friction between the ladder and the ground.

3 A uniform beam has length 4 metres. One end rests on horizontal ground, and the beam is in equilibrium at 10° to the horizontal, propped against a smooth marble slab of height 0.4 metres, as shown in the diagram. Find the least possible coefficient of friction between the beam and the ground.

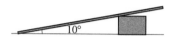

4 Use a geometrical method to rework Exercise 4A Question 4.

5 Use a geometrical method to rework Exercise 4B Question 3.

6 Use a geometrical method to rework Exercise 4B Question 4.

7 Use a geometrical method to rework Exercise 4B Question 6.

Miscellaneous exercise 7

You should use a geometrical method to work the problems in this exercise.

1 $ABCD$ is a square table of side 3 metres. Sipho and Tandi push it together. Sipho pushes at A in the direction of the edge AB with a force of 60 newtons. Tandi pushes at the midpoint of the edge CD perpendicular to that edge with a force of 80 newtons. Where would one person have to push it, and with what force, to produce the same total effect as Sipho and Tandi?

2 A tree trunk is to be moved by a resultant force through its centre of mass at 55° to its length. It is pulled by cables from two tractors, with the same tension in each cable. The cable from one tractor is attached at a point 5 metres from the centre of mass, and this pulls at 70° to the trunk. Where should the other cable be attached, and at what angle to the trunk should this be pulled?

3 Two parallel horizontal rails are 25 cm apart, at the same level. A uniform box with rectangular cross-section 34 cm × 22 cm is balanced on the rails at an angle of $\theta°$ to the horizontal. If the contact of the faces of the box with the rails is smooth, show that for the box to rest in equilibrium

$$\frac{25\cos\theta° - 17}{25\sin\theta° - 11} = \tan\theta°$$

Deduce that $25\cos 2\theta° = 17\cos\theta° - 11\sin\theta°$, and use a numerical method (P2&3 Chapter 8) to estimate the value of θ.

4 An athlete is training in the gym. In one exercise he stands upright with his chest pressed against a horizontal rail at shoulder level. He grasps the rail and then leans back with his body straight until his arms are fully stretched. Estimates of his physical measurements are:

Height of shoulders above the floor	150 cm
Height of centre of mass above the floor	120 cm
Length of outstretched arms	60 cm

Use a simplified model to estimate the least possible coefficient of friction between his feet and the floor if he is not to slip while doing this exercise.

5 i A uniform ladder is leaning against a smooth vertical wall and stands on a rough horizontal floor without slipping. The base of the ladder is 1 m from the wall, and the top of the ladder is 3 m from the floor. Show that the coefficient of friction, m, between the ladder and the floor satisfies $\mu \geq \frac{1}{6}$.

ii A man, whose weight is three times that of the ladder, climbs to a point three-quarters of the way up the ladder. Find the minimum possible value of μ if the ladder does not slip.

6 A uniform rod has length 2 m and mass 4 kg. It is held in equilibrium by a string attached to one end, while the other end rests on rough horizontal ground. The rod makes an angle of 30° with the horizontal and the string makes an angle of $\theta°$ with the vertical (see diagram). The coefficient of friction between the rod and the ground is $\dfrac{1}{\sqrt{3}}$ and the rod is on the point of slipping. Determine θ.

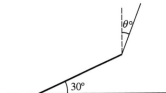

7 For the situation in Miscellaneous exercise 4 Question 8, show that the centre of mass of the prism is a horizontal distance 1.7a from A. Hence obtain the condition $\tan \beta < \dfrac{46}{17}$ for the prism to slide before it starts to lift.

8 For the situation of Miscellaneous exercise 4 Question 5, use a geometrical method to show that the tension in the string is 960 N, and hence determine the total force exerted by the hinge on the beam.

Chapter 8
Centres of mass of special shapes

In Chapter 3 all the objects are made up from components whose centres of mass are at their geometrical centres. In this chapter the ideas are extended to various shapes which do not have a geometrical centre. When you have completed the chapter, you should

- know formulae for locating the centre of mass of objects modelled as wires, laminas, solids or shells of various standard shapes
- be able to find the centre of mass of objects made up of components with these shapes
- be able to find the centre of mass of objects formed by removing parts from an object whose centre of mass you already know.

At this stage it is not possible to give reasons for many of the results stated. For most of the shapes considered in this chapter, and many others, finding the centre of mass involves methods using integration.

A summary of the results in this chapter is given in the formulae list at the back of this book.

8.1 Uniform wire shapes

Here 'wire' is used to mean an object which can be modelled by a curve having no thickness. A uniform wire has constant mass per unit length.

You already know how to find the centre of mass of a wire bent into straight sections, such as the triangle in Example 3.2.2. Another important shape is a circular arc.

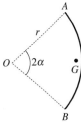

Fig. 8.1 shows a wire AB bent into the shape of an arc of a circle of radius r, centre O. The centre of mass G is obviously on the radius which bisects the angle AOB, so it is convenient to denote the angle AOB by 2α, so angle $AOG = \alpha$. Then, *if the angle α is measured in radians*, it can be shown that

Fig. 8.1

$$OG = \frac{r\sin\alpha}{\alpha}$$

An important special case is when the wire is bent into a semicircle. Then angle $AOB = \pi$, so $\alpha = \frac{1}{2}\pi$ and $\sin\alpha = 1$, giving

$$OG = \frac{2r}{\pi}$$

Although the general formula can't be proved at this stage, here are two arguments which suggest that it is reasonable. (For a third argument, see Exercise 8A Question 12.)

- It is shown in P2&3 Section 6.1 that if $0 < \theta < \frac{1}{2}\pi$, $\cos\theta < \dfrac{\sin\theta}{\theta} < 1$. It is easy to see that this in fact holds if $0 < \theta < \pi$. Writing θ as α and multiplying by r, this gives $r\cos\alpha < OG < r$. So, as you would expect, the formula gives a position for G which always lies inside the 'bow' formed by the arc AB and the chord AB.

- If you put together two arcs like Fig. 8.1, with centres of mass G_1 and G_2, as in Fig. 8.2, then you get an arc making an angle 4α at O. Its centre of mass must be at G, the mid-point of G_1G_2. The distance OG is therefore $OG_1\cos\alpha$, so

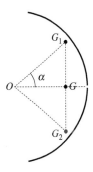

$$OG = \frac{r\sin\alpha\,\cos\alpha}{\alpha}$$

$$= \frac{r\times\sin\alpha\,\cos\alpha}{2\alpha} = \frac{r\sin 2\alpha}{2\alpha}$$

Fig. 8.2

This is what you would expect from the formula, replacing α by 2α.

EXAMPLE 8.1.1

A letter D is formed by bending a uniform wire into the shape of a semicircle and its diameter. If this hangs freely from its top left corner, what will be the angle between the upright and the vertical?

Denote the radius of the semicircle by r, and suppose that the wire has mass k per unit length. Let G be the centre of mass of the letter D, O the mid-point of the diameter, and let $OG = \bar{x}$. Then the data are summarised in Table 8.3.

Table 8.3

Mass	$2rk$	πrk	$(2 + \pi)rk$
Distance from O	0	$\dfrac{2r}{\pi}$	$\bar{x}$

The usual formula (from the box on page 42) gives

$$\bar{x} = \frac{2kr \times 0 + \pi rk \times \left(\dfrac{2r}{\pi}\right)}{(2 + \pi)rk} = \frac{2r^2k}{(2 + \pi)rk} = \frac{2r}{2 + \pi}.$$

Fig. 8.4 shows the letter hung from the top left corner, with the vertical joining that corner to the centre of mass. The upright makes an angle $\tan^{-1} \dfrac{2}{2 + \pi}$ with the vertical, which is about 21°.

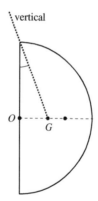

Fig. 8.4

8.2 Uniform lamina shapes

Here 'lamina' is used to mean an object which can be modelled by a plane region with no thickness. A uniform lamina has constant mass per unit area.

You already know that if a uniform lamina has a point of central symmetry then that point is the centre of mass.

A triangle doesn't have a centre of symmetry. One way to find its centre of mass is to cut it into a large number of narrow strips parallel to one side, as in Fig. 8.5. To a close approximation the centre of mass of each strip lies on the median, that is the line which joins the mid-point of

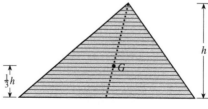

Fig. 8.5

that side to the opposite vertex. So the centre of mass of the triangular lamina lies somewhere on this line.

You can use this argument for each of the three sides in turn. It follows that the centre of mass is at the point where the three medians meet, which is called the centroid of the triangle. (See P1 Example 13.5.2, where it is proved that, if the vertices have position vectors $\mathbf{a}$, $\mathbf{b}$ and $\mathbf{c}$, then the medians meet at the point G with position vector $\frac{1}{3}(\mathbf{a}+\mathbf{b}+\mathbf{c})$.) An important property of the centroid is that, along each median, it lies one-third of the way from the mid-point of the side to the vertex.

EXAMPLE 8.2.1

A uniform trapezium-shaped lamina has its vertices at the origin O and at the points $A(8,6)$, $B(28,6)$ and $C(30,0)$. Find the coordinates $(\bar{x},\bar{y})$ of its centre of mass. (See Fig. 8.6.)

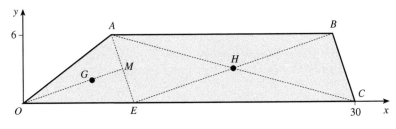

Fig. 8.6

If E is the point $(10,0)$, think of the trapezium as made up of the triangle OEA and the parallelogram $ECBA$. Their areas (in appropriate units) are $\frac{1}{2}\times 10\times 6=30$ and $20\times 6=120$. Suppose that the mass of a unit area of the lamina is k.

The centre of mass of the triangle is at G, two-thirds of the way along the median OM, where M is the mid-point of AE. M has coordinates $\left(\frac{1}{2}(8+10),\frac{1}{2}(6+0)\right)$ which is $(9,3)$, so G has coordinates $(6,2)$.

The parallelogram has point symmetry about the point H where its diagonals intersect; for every point P inside the parallelogram, there is another point Q inside the parallelogram such that H is the mid-point of PQ. The centre of mass of the parallelogram is therefore at H, whose coordinates are $(19,3)$.

The data are summarised in Table 8.7.

Table 8.7

	Triangle	Parallelogram	Trapezium
Mass	$30k$	$120k$	$150k$
x-coordinate	6	19	$\bar{x}$
y-coordinate	2	3	$\bar{y}$

The usual equations give

$$\bar{x} = \frac{30k \times 6 + 120k \times 19}{150k} = 16.4, \quad \bar{y} = \frac{30k \times 2 + 120k \times 3}{150k} = 2.8.$$

The centre of mass of the trapezium is at the point $(16.4, 2.8)$.

Another lamina shape with no centre of symmetry is a sector of a circle, shown as OAB in Fig. 8.8. Its centre of mass can also be located by cutting it into a large number of pieces, but this time along radii. Each piece is then very nearly a triangle, with its centre of mass $\frac{2}{3}r$ from O to a very close approximation. These centres of mass lie on a circular arc of radius $\frac{2}{3}r$. So if G is the centre of mass of the sector, the formula in Section 8.1 gives

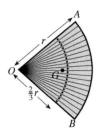

$$OG = \frac{2r \sin \alpha}{3\alpha}$$

Fig. 8.8

where the angle AOB is 2α radians.

An important special case is the semicircular lamina, for which $\alpha = \frac{1}{2}\pi$. The formula then gives

$$OG = \frac{4r}{3\pi}.$$

EXAMPLE 8.2.2

A uniform metal plate, in the shape of a sector OAB of a circle centre O, is free to rotate about a horizontal axis through A and B. What can you say about the angle AOB if the plate can swing like a pendulum with O below AB?

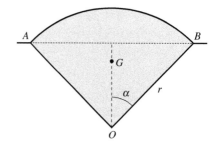

Denote the radius of the circle by r, the angle AOB by 2α radians, and the centre of mass of the plate by G (see Fig. 8.9).

Fig. 8.9

For the plate to swing like a pendulum with O below AB, the centre of mass must be below AB. Then, if the plate is displaced from the vertical position by a small angle, the moment of the weight will act so as to restore the plate to the vertical. The condition for this to occur is that OG is less than the distance from O to AB, so $\dfrac{2r \sin \alpha}{3\alpha} < r \cos \alpha.$

Dividing both sides by r and using $\dfrac{\sin \alpha}{\cos \alpha} = \tan \alpha$, this inequality can be rearranged as

$$\tan \alpha < \tfrac{3}{2}\alpha.$$

This cannot be solved exactly, but you can use graphs or a numerical method to find an approximate solution. Fig. 8.10 shows the graphs of $y = \tan x$ and $y = \frac{3}{2}x$, which intersect where $x \approx 0.967$. You can see that $\tan \alpha < \frac{3}{2}\alpha$ if α is less than this value. So the plate will swing with O below AB if angle $A\hat{O}B$ is less than 2×0.967 radians, which is about $111°$.

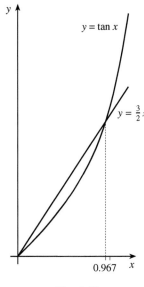

Fig. 8.10

Exercise 8A

1 A uniform lamina $ABCDE$ is made in the shape of a trapezium, as shown in the diagram. $AE = 10$ cm, $ED = 6$ cm and AE and BD are both perpendicular to AC. The centre of mass of the lamina lies on the line BD. Find the length of AC.

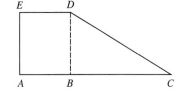

2 A uniform rigid wire AB is in the form of a circular arc of radius 1.5 m with centre O. The angle AOB is a right angle. The wire is in equilibrium, freely suspended from the end A.
The chord AB makes an angle of $\theta°$ with the vertical (see diagram).

   i Show that the distance of the centre of mass of the arc from O is 1.35 m, correct to 3 significant figures.

   ii Find the value of θ.

   (Cambridge International AS and A Level Mathematics 9709/05 Paper 5 Q2
   June 2008)

3 A uniform lamina is in the form of a sector of a circle with centre O, radius 0.2 m and angle 1.5 radians. The lamina rotates in a horizontal plane about a fixed vertical axis through O. The centre of mass of the lamina moves with speed 0.4 m s^{-1}. Show that the angular speed of the lamina is 3.30 rad s^{-1}, correct to 3 significant figures.

   (Cambridge International AS and A Level Mathematics 9709/05 Paper 5 Q1
   June 2009)

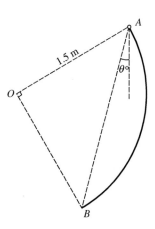

4 A uniform lamina *ABCD* is in the form of a trapezium in which *AB* and *DC* are parallel and have lengths 2 m and 3 m respectively. *BD* is perpendicular to the parallel sides and has length 1 m (see diagram).

i Find the distance of the centre of mass of the lamina from *BD*.

The lamina has weight *W* N and is in equilibrium, suspended by a vertical string attached to the lamina at *B*. The lamina rests on a vertical support at *C*. The lamina is in a vertical plane with *AB* and *DC* horizontal.

ii Find, in terms of *W*, the tension in the string and the magnitude of the force exerted on the lamina at *C*.

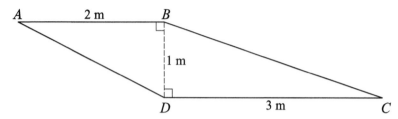

(Cambridge International AS and A Level Mathematics 9709/05 Paper 5 Q3 November 2005)

5 A kite is constructed of a thin uniform sheet. It is symmetrical about its longer diagonal, whose length is 1.2 metres, and the diagonals cross 0.45 metres from one end of it. Find the distance of the centre of mass from the point where the diagonals cross.

6 A quadrilateral has vertices at the points with coordinates (0,0), (12,0), (9,12) and (0,9). Find the coordinates of the centre of mass of a uniform lamina bounded by this quadrilateral.

7 A mirror glass has the shape of a rectangle, of width *w* and height *h*, surmounted by a semicircle of diameter *w*. Show that the height of the centre of mass above the base of the rectangle is $\dfrac{12h^2 + 3\pi hw + 2w^2}{24h + 3\pi w}$.

8 A trough has a cross-section in the form of a trapezium. Its base has length 1 metre, and the sides slope out at 45° to the horizontal. The trough is filled with feed to a depth of *x* metres. Find the value of *x* given that the centre of mass of the contents of the trough is $\frac{1}{2}$ metre above the base.

9 A strip light is shaped into the outline of a crescent with pointed corners at *A* and *B*, 2 metres apart. The convex edge of the crescent is a semicircle; the concave edge is one-sixth of a circle of radius 2 metres. Find the distance of the centre of mass of the crescent from the line *AB*.

10 A trapezium-shaped block of uniform thickness has the corners of one of its faces labelled *A*, *B*, *C* and *D*, as shown in the diagram. Angles *ABC* and *BCD* are right angles. The block stands on a plane which is inclined at an angle *α* to the horizontal. In which of the following configurations P, Q, R and S is the block on the point of toppling when tan *a* is

a $\dfrac{11}{24}$ **b** $\dfrac{19}{21}$ **c** $\dfrac{24}{19}$ **d** $\dfrac{89}{57}$?

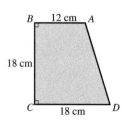

P *AB* coincides with a line of greatest slope with *B* above *A*.

Q *BC* coincides with a line of greatest slope with *C* above *B*.

R *CD* coincides with a line of greatest slope with *D* above *C*.

S *DA* coincides with a line of greatest slope with *A* above *D*.

11 A structure made from uniform wire consists of the arc *AB* of a circle of radius 10 cm, which subtends an angle of 110° at its centre, together with the chord *AB*. The structure is suspended freely from *A*, as shown in the diagram. Find the angle $\theta°$ which the chord *AB* makes with the downward vertical.

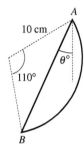

12 The diagram shows a uniform circular wire of radius *r* whose centre is *O*. It is cut at *A* and *B* into two parts, where angle *AOB* = 2α. If its mass per unit length is *k*, write down the mass of each part and the distance of its centre of mass from *O*.

Use the formula $\bar{x} = \dfrac{m_1 x_1 + m_2 x_2}{m_1 + m_2}$ to verify that the centre of mass of the two parts together is at *O*.

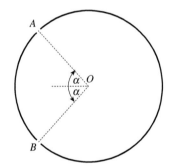

13 A uniform lamina is a quadrant of a circle with *OA*, *OB* as its bounding radii. It hangs freely from *A*. Show that *OA* makes an angle $\tan^{-1} \dfrac{4}{3\pi - 4}$ with the vertical.

8.3 Uniform solid shapes

A uniform solid has constant mass per unit volume.

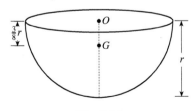

Fig. 8.11

Fig. 8.11 shows a solid hemisphere with centre *O* and radius *r*. Since this is a solid of revolution (see P1 Chapter 17), its centre of mass *G* lies on the axis of rotation. It can be proved that $OG = \frac{3}{8}r$.

Fig. 8.12 shows a circular cone, with its vertex V at a height h above the centre of the base. Its centre of mass lies on the axis of rotation at a height $\frac{1}{4}h$ above the base.

The result for the cone can be generalised to any uniform solid cone or pyramid. If C is the centre of mass of the base, considered as a lamina, and if the vertex is V, then the centre of mass lies one-quarter of the way up the line joining C to V.

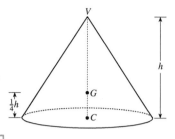

Fig. 8.12

EXAMPLE 8.3.1

Fig. 8.13 shows a toy made of a solid cone, of height h and base radius r, and a solid hemisphere of radius r, glued together across their flat surfaces. The toy stands on a rough horizontal floor with the vertex of the cone pointing upwards. If it is given a small knock sideways, what will happen?

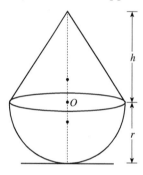

Fig. 8.13

In the figure O is the centre of the circle common to the cone and the hemisphere. Let $\overline{y}$ be the height of the centre of mass above O. Suppose that the toy is made of material of mass k per unit volume. The data are in Table 8.14.

Table 8.14

	Cone	Hemisphere	Whole toy
Mass	$\frac{1}{3}\pi r^2 h k$	$\frac{2}{3}\pi r^3 k$	$\frac{1}{3}\pi r^2(h+2r)k$
Height above O	$\frac{1}{4}h$	$-\frac{3}{8}r$	$\overline{y}$

This gives

$$\overline{y} = \frac{\frac{1}{12}\pi r^2 h^2 k - \frac{1}{4}\pi r^4 k}{\frac{1}{3}\pi r^2(h+2r)k} = \frac{\frac{1}{4}(h^2 - 3r^2)}{(h+2r)}.$$

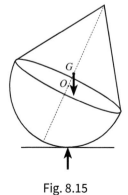

Fig. 8.15

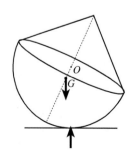

Fig. 8.16

What happens after the toy is knocked depends on whether $\bar{y} > 0$ or $\bar{y} < 0$.

Fig. 8.15 shows the situation $\bar{y} > 0$, so that the centre of mass is above O.

Then the weight of the toy has a clockwise moment about the point of contact of the toy and the floor, so the toy will fall over further.

If $\bar{y} < 0$, as in Fig. 8.16, the weight has an anticlockwise moment, which returns the toy to its vertical position.

So if $h > \sqrt{3}r$, the toy will fall over; if $h < \sqrt{3}r$, it will stay upright.

In this example the vertical position of the toy is always a position of equilibrium. If a small knock makes it fall over further, the equilibrium is called **unstable**; if it returns towards the equilibrium position, the equilibrium is **stable**.

8.4 Uniform shell shapes

Here 'shell' is used to mean an object which can be modelled by a curved surface having no thickness. For example, the peel of an orange or the earth's crust could be modelled as a spherical shell. A uniform shell has constant mass per unit area.

The centre of mass of any shell which is a surface of revolution has its centre of mass on the axis of rotation. For a hemispherical shell with centre O and radius r, the centre of mass is $\frac{1}{2}r$ from O (see Fig. 8.17).

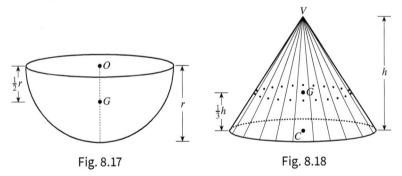

Fig. 8.17   Fig. 8.18

Here 'conical shell' is used to mean simply the curved surface of a cone, not including the base. Like the circular sector in Fig. 8.8, this can be cut up into a large number of near-triangular strips, each of which would have its centre of mass $\frac{1}{3}h$ above the base, as illustrated in Fig. 8.18. All these centres of mass would lie on a circle. It follows that the centre of mass of the complete conical shell is $\frac{1}{3}h$ above the base.

EXAMPLE 8.4.1

Fig. 8.19 shows a lantern made of glass of uniform thickness. It is formed by joining together a conical shell of height 6 cm, a cylindrical shell of height 10 cm and a hemispherical shell, all of radius 8 cm. Find the depth of the centre of mass of the lantern below the vertex V of the cone.

Let the mass of the glass be k kg for each cm² of its surface area.

The area of a conical surface is given by the formula $\pi r l$, where l is the slant height. In this case the slant height is $\sqrt{6^2 + 8^2}\,\text{cm} = 10\text{cm}$, so the surface area is $\pi \times 8 \times 10$ cm. The area of the cylindrical surface is $2\pi \times 8 \times 10$ cm², and the area of the hemispherical surface is $2\pi \times 8^2$ cm².

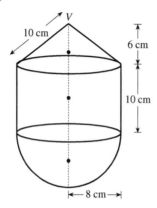

Fig. 8.19

Using the results for a conical shell and a hemispherical shell given above, the centre of mass of the cone is $\frac{1}{3} \times 6$ cm above its base, and the centre of mass of the hemisphere is $\frac{1}{2} \times 8$ cm below its centre. The centre of mass of the cylinder is at its geometrical centre.

The data are summarised in Table 8.20.

Table 8.20

	Cone	Cylinder	Hemisphere
Mass (kg)	$80\pi k$	$160\pi k$	$128\pi k$
Depth below V (cm)	4	11	20

The distance of the centre of mass below V, in cm, is therefore

$$\frac{80\pi k \times 4 + 160\pi k \times 11 + 128\pi k \times 20}{80\pi k + 160\pi k + 128\pi k} \approx 12.6.$$

The centre of mass of the lantern is about 12.6 cm below V.

8.5 Finding centres of mass by subtraction

Sometimes an object is formed not by putting several parts together, but by removing bits from an object which was originally whole. The examples in this section show how to find the centre of mass in such cases.

EXAMPLE 8.5.1

A mechanism includes a uniform circular metal plate of radius 10 cm, with two circular holes cut out of it. The design specification describes these by marking two diameters as coordinate axes on the plate. One hole has radius 2 cm with its centre at $(6, 0)$, the other has radius 4 cm with its centre at $(-2, 5)$. Find the centre of mass of the plate with holes in it, shown in Fig. 8.21.

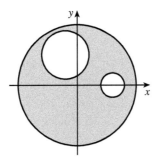

Fig. 8.21

If the mass of the metal is k kg per cm^2, the original plate had mass $100\pi k$ kg, and the metal removed from the holes had mass $4\pi k$ kg and $16\pi k$ kg. The metal that remains therefore has mass $80\pi k$ kg.

You can argue that, if you put the metal back in the two holes, then the three parts would make up the complete circular plate that you started with. This could be shown by the data in Table 8.22.

Table 8.22

Mass (kg)	$80\pi k$	$4\pi k$	$16\pi k$	$100\pi k$
x-coordinate (cm)	$\bar{x}$	6	−2	0
y-coordinate (cm)	$\bar{y}$	0	5	0

From this you can make up the equations (cancelling the common factor πk)

$$80\bar{x} + 4 \times 6 + 16 \times (-2) = 100 \times 0 \text{ and } 80\bar{y} + 4 \times 0 + 16 \times 5 = 100 \times 0.$$

These give $\bar{x} = \dfrac{8}{80} = 0.1$ and $\bar{y} = -1$.

The centre of mass of the plate with holes in it is at $(0.1, -1)$.

EXAMPLE 8.5.2

In Fig. 8.23, OAB is a quadrant of a circle made of plywood. The triangle OAB is sawn off and discarded, to leave a segment bounded by the line AB and the arc AB. The line AB has length 2 metres. Find the distance of the centre of mass G of the segment from the line AB.

You can make an equation by expressing the fact that the triangle and the segment together make up the quadrant OAB.

If M is the mid-point of AB, the distance OM is 1 m so the triangle OAB has area 1 m^2 and its centre of mass is $\frac{2}{3}$ m from O.

The circle has radius $\sqrt{2}$ m, so the original quadrant has area $\frac{1}{4}\pi \times \left(\sqrt{2}\right)^2$ m^2, which is $\frac{1}{2}\pi$ m^2. The distance of its centre of mass from O is given by the

formula $\dfrac{2r\sin\alpha}{3\alpha}$ with $\alpha = \frac{1}{4}\pi$, so this is $\dfrac{2\sqrt{2} \times \dfrac{1}{\sqrt{2}}}{3 \times \frac{1}{4}\pi}$ simplified to give $\dfrac{8}{3\pi}$ m.

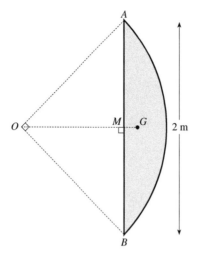

Fig. 8.23

Suppose that the plywood has mass k kg per unit area. Denote OG by $\bar{x}$ metres. Then the data can be summarised as in Table 8.24.

Table 8.24

	Triangle	Segment	Quadrant
Mass (kg)	k	$\left(\frac{1}{2}\pi - 1\right)k$	$\frac{1}{2}\pi k$
Distance from O (m)	$\frac{2}{3}$	$\bar{x}$	$\dfrac{8}{3\pi}$

So $\bar{x}$ can be found from the equation

$$\tfrac{2}{3}k + \left(\tfrac{1}{2}\pi - 1\right)k\bar{x} = \tfrac{1}{2}\pi k \times \frac{8}{3\pi}, \text{ giving } \bar{x} = \frac{\frac{2}{3}}{\frac{1}{2}\pi - 1} \approx 1.168$$

G is about 1.168 metres from O, which is 0.168 metres from AB.

Exercise 8B

1 A uniform solid consists of a cylindrical part of radius r and length l, and a hemispherical part of radius r. One end of the cylinder is in contact with the plane face of the hemisphere, and there is no overlap. The solid is held on a horizontal surface, with a generator AB of the cylinder in contact with the horizontal surface, as shown in the diagram. The solid is then released. Determine whether the cylindrical part continues to be in contact with the horizontal surface, or whether the solid rolls so that the hemisphere is in contact with the horizontal surface, when

a $l = r$,   **b** $l = \tfrac{1}{2}r$.

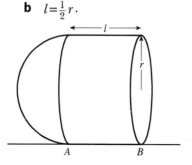

2 A uniform square of cardboard has sides of length 20 cm. A square hole is punched in it, with its centre 4 cm from one side and 3 cm from an adjacent side. The sides of the hole have length 4 cm. Find the position of the centre of mass of the card that remains. Does the answer depend on the angle at which the hole is punched?

3 A uniform solid cone, of radius 10 cm and height 16 cm, has a circular hole of radius 5 cm bored into its base to a depth of 6 cm. The axis of the hole coincides with the axis of the cone. The cone is freely suspended from a point of the circumference of the base of the cone. Find α, the acute angle that the axis makes with the vertical.

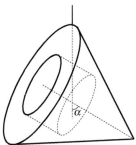

4 A closed container consists of a hemispherical shell of mass 0.04 kg and radius 3 cm, and a conical shell of mass 0.045 kg, radius 3 cm and height 6 cm. The shells are joined so that their axes of symmetry coincide, as shown in both figures. The container holds 18π cm³ of fine dry sand of density 2.3 kg per 1000 cm³. The container is held with its axis vertical, firstly with the conical part uppermost as in the first figure, then with the hemispherical part uppermost, as in the second figure. Show that the sand just occupies the hemispherical part in the first case, and just occupies the conical part in the second case. Find, in each case, the height of the centre of mass of the container (with sand), above its lowest point.

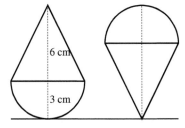

5 The wing of a hang-glider is a uniform lamina, formed by removing from a square of side l a quadrant of a circle of radius l, with its centre at one corner of the square. Find the distance of the centre of mass of the wing from the opposite corner.

6 A traffic cone consists of a 45 cm × 45 cm square base of height 5 cm, and a conical shell of radius 15 cm and height 75 cm. The base has a circular hole through it, of radius 15 cm, to facilitate stacking. The base is made of rubber of density 1 kg per 1000 cm³ and the conical shell has mass 0.5 kg. The cone is held with an edge of the base in contact with a horizontal surface, and with the axis of the shell making an angle α with the horizontal, as shown in the diagram. The cone is now released. Find the angle β such that the cone assumes the upright position if $\alpha > \beta$ and the cone topples if $\alpha < \beta$.

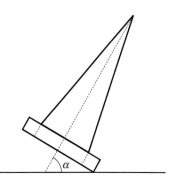

7 A 16 cm × 6 cm rectangular hole is cut in a uniform circular lamina of radius 12 cm. One of the 16 cm sides of the rectangle, labelled AB, coincides with a diameter of the circle, and the mid-point of AB coincides with O, the centre of the circle.

The lamina is freely suspended from a point at the end of the diameter containing AB, as shown in the diagram. Find the angle θ between this diameter and the downward vertical.

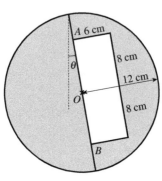

Miscellaneous exercise 8

1 A right-angled triangular prism has weight W. The sides containing the right angle have lengths $3a$ and b. The prism is at rest with its edges of length $3a$ in contact with a horizontal table. When a horizontal force of magnitude P is applied to the mid-point of the uppermost edge, at right angles to the vertical face, the prism topples if the direction of the force is as shown in the first diagram and slides if the direction is as shown in the second diagram. Show that $\dfrac{a}{b} < \mu < \dfrac{2a}{b}$, where μ is the coefficient of friction between the prism and the table.

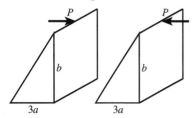

2 A uniform solid cone with height 0.8 m and semi-vertical angle 30° has weight 20 N. The cone rests in equilibrium with a single point P of its base in contact with a rough horizontal surface, and its vertex V vertically above P. Equilibrium is maintained by a force of magnitude F N acting along the axis of symmetry of the cone and applied to V (see diagram).

i Show that the moment of the weight of the cone about P is 6 N m.

ii Hence find F.

(Cambridge International AS and A Level Mathematics 9709/51 Paper 5 Q2 November 2014)

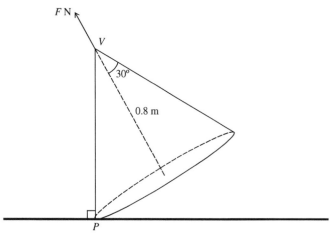

3 A conical shell has weight W and slant height equal to 4 times the radius of its rim. The shell rests with its surface in contact with horizontal ground. A force of magnitude T is applied to the shell's vertex O in a direction making an angle β with the ground, as shown in the diagram.

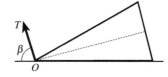

a Given that O is on the point of leaving the ground, show that $T\sin\beta = \frac{3}{8}W$.

b The coefficient of friction between the shell's surface and the ground is 0.4, and O is on the point of moving along the ground, show that
$$T(5\cos\beta + 2\sin\beta) = 2W.$$

Deduce that if T is gradually increased from zero the shell will start to lift at O before sliding if $\tan\beta > \frac{3}{2}$.

4 Repeat Question 3 when the force of magnitude T is applied at the point P of the rim which is in contact with the ground (as shown in the diagram), showing that $T\sin\beta = \frac{5}{8}W$ when P is on the point of leaving the ground, $T(5\cos\beta + 2\sin\beta) = 2W$ as before in part **b**, and $\tan\beta > \frac{25}{6}$ in the final part.

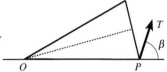

5 A vessel consists of a base, two ends and two sides, all of the same uniform material, arranged as shown in the diagram. The base is a 12 cm × 5 cm rectangle, and the ends are 8 cm × 5 cm and 10 cm × 5 cm rectangles.

Each of the sides is in the form of a trapezium $ABCD$ in which $AB = 12$ cm, $BC = 10$ cm, $CD = 18$ cm and $DA = 8$ cm, and which is right-angled at A and at D. Show that the centre of mass of the vessel is about 7.52 cm from the end containing DA. Deduce that the vessel can stand on its base without toppling.

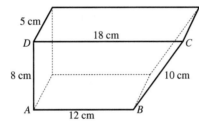

6 A vessel consists of a base, two ends and two sides, arranged as shown in the diagram. The base is a 6 cm × 4 cm rectangle, and the ends are 5 cm × 4 cm and 13 cm × 4 cm rectangles. Each of the sides is in the form of a trapezium $ABCD$ in which $AB = 6$ cm, $BC = 13$ cm, $CD = 18$ cm and $DA = 5$ cm, and which is right-angled at A and at D. The base is sufficiently heavy for the vessel to be capable of standing on its base when empty, without toppling. Water is poured slowly into the vessel. Assuming the weight of the vessel can be ignored, show that the vessel topples when the depth of water reaches d cm, where $d^2 = 18.75$.

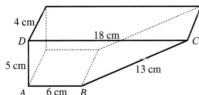

7 A uniform lamina of weight W has the shape of a semicircle with diameter AB. The point A is pivoted at a fixed point. The lamina is kept in equilibrium, with AB vertical, by a force of magnitude P acting horizontally with a line of action passing through the centre of the semicircle.

Find the magnitude and direction of the force exerted by the pivot on the lamina at A.

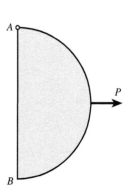

8 A uniform wire, in the shape of a semicircle, is at rest on a smooth horizontal table with its diameter making an angle α with the horizontal $\left(0 < \alpha < \tan^{-1} \frac{1}{2}\pi\right)$. The wire is held in this position by a force of magnitude P acting vertically upwards at the uppermost point of the wire. The force exerted by the table on the hemisphere is R (see diagram).

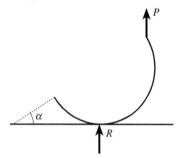

Show that $\dfrac{R}{P} = \dfrac{\pi}{2\tan\alpha} - 1$.

9 A light vase has a 12 cm × 12 cm square cross-section and height 30 cm. The vase is placed on a plane inclined at $\tan^{-1} \frac{1}{2}$ to the horizontal, with two parallel edges of its base coinciding with lines of greatest slope of the plane. Water is poured slowly into the vase. Determine whether the vase topples without first overflowing or whether it overflows without toppling.

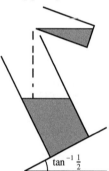

10 A uniform solid consisting of two parts, a circular cylinder of radius 6 cm and height 10 cm and a cone frustum of end radii 6 cm and 2 cm and height 10 cm, is arranged as shown in the diagram. Find the distance of the centre of mass of the solid from its base.

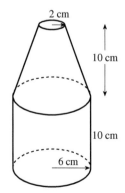

11 A sector of a circle of radius 10 cm has angle 2.5 radians. Find the distance of the centre of mass of the sector from the centre of the circle by

   a direct application of the relevant formula in Section 8.2,

   b treating the sector as a circle from which the corresponding major sector has been removed.

12 A uniform rectangular lamina, L cm $\times$ 10 cm, has a circular hole in it of radius 4 cm. The centre of the hole lies on the line midway between the sides of length L cm, and the line midway between the sides of length 10 cm is a tangent to the circular boundary of the hole. When suspended freely from a corner as shown in the diagram, the sides of length L cm make an angle of $25°$ with the vertical. Find L.

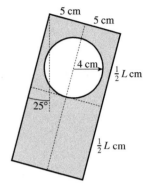

13 A uniform lamina is made up of two rectangular parts, each 10 cm $\times$ 20 cm, and a part in the form of a circular sector with centre at O, radius 10 cm and angle $130°$, as shown in the diagram. The centre of mass of the lamina is at G. Find the distance OG.

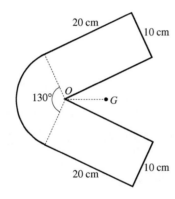

Chapter 9
Linear motion with variable forces

This chapter is about the motion of a particle along a straight line when the forces acting on it depend on time or on its velocity or position. When you have completed the chapter, you should

- know how to describe the motion by a differential equation involving two or more of the quantities displacement, velocity and time
- be able to solve the differential equation in cases where the variables are separable, and interpret the solution
- know that acceleration can be expressed as $v\dfrac{\mathrm{d}v}{\mathrm{d}x}$
- know that the work done by a force can be expressed as $\int F\,\mathrm{d}x$.

To use this chapter you need to be familiar with the methods of solving differential equations described in P2&3 Chapter 19. Some sections require knowledge of integration using partial fractions or the method of substitution (P2&3 Chapters 15 and 18).

9.1 Velocity–time equations

When an object is moving in a straight line under the action of constant forces, you know from Newton's second law that the acceleration is constant. You can then use the standard methods for constant acceleration to get equations relating the velocity, displacement and time.

But often the forces on an object are not constant. For example, if you push a table across the floor, you may push hard to get it started and then reduce the force once the table is moving. On the other hand, if you are driving a car, you may apply a light touch on the accelerator as you move off from rest and then increase the pressure as the car gains speed. In such cases the acceleration will not be constant.

You know, however, from M1 Section 11.1, that the acceleration can be written as $\dfrac{\mathrm{d}v}{\mathrm{d}t}$.

So if you have a formula for the force in terms of the time, you can express the acceleration as a function of the time, and then integrate this to find an expression for the velocity.

EXAMPLE 9.1.1

A small car of mass 800 kg is powered by an electric motor. The driver applies a gradually increasing accelerating force, given by $60\sqrt{t}$ newtons after t seconds. How long does the car take to reach its maximum speed of 10 m s^{-1}?

By Newton's second law the acceleration of the car is $\dfrac{60}{800}\sqrt{t}$. So if the velocity after t seconds is v m s^{-1}, the motion is described by the differential equation

$$\frac{\mathrm{d}v}{\mathrm{d}t} = \frac{3}{40}\sqrt{t}.$$

This can be integrated directly to give

$$v = \tfrac{1}{20}t^{\frac{3}{2}} + k.$$

Since t denotes the time after the car starts to move, $v = 0$ when $t = 0$, so $k = 0$. The (t,v) equation is therefore

$$v = \tfrac{1}{20}t^{\frac{3}{2}}.$$

You are told that the maximum speed is 10 m s^{-1}, so the (t,v) equation holds for velocities up to $v = 10$. When $v = 10$, $t^{\frac{3}{2}} = 200$, so $t = 200^{\frac{2}{3}} = 34.19\ldots$.

The small car takes 34.2 seconds to reach its maximum speed.

Variable acceleration does not always appear as a function of t. Another cause of variation is that the force may depend on how fast an object is moving. One example is air resistance, which increases as speed increases. Another is when an engine is working at constant power, in which case the driving force decreases as the speed increases.

Such examples lead to differential equations that give $\dfrac{dv}{dt}$ in terms of v. These can be solved by the method described in P2&3 Section 19.3, using the property that

$$\frac{dv}{dt} \times \frac{dt}{dv} = 1.$$

A differential equation of the form $\dfrac{dv}{dt} = f(v)$ can then be written as $\dfrac{dt}{dv} = \dfrac{1}{f(v)}$, and this can then be integrated directly.

EXAMPLE 9.1.2

A cyclist and her bicycle have total mass 100 kg. She is working at constant power of 80 watts. Calculate how long it takes her to accelerate from 4 m s^{-1} to 8 m s^{-1} along a level road,

a if air resistance is neglected,

b making allowance for air resistance of $0.8v$ newtons when her speed is $v \text{ m s}^{-1}$.

a Power can be calculated as force × velocity, so at $v \text{ m s}^{-1}$ her efforts produce a driving force of $\dfrac{80}{v}$ newtons. Writing her acceleration as $\dfrac{dv}{dt}$, Newton's second law gives

$$\frac{80}{v} = 100\frac{dv}{dt}.$$

To solve this differential equation, write

$$\frac{dt}{dv} = 1 \Big/ \frac{dv}{dt} = 1 \Big/ \frac{4}{5v} = \frac{5}{4}v.$$

This can be integrated with respect to v, to give

$$t = \tfrac{5}{8}v^2 + k.$$

Suppose that time is measured from the instant when the cyclist's speed is 4 m s^{-1}. Then $t = 0$ when $v = 4$, so

$$0 = \tfrac{5}{8} \times 4^2 + k, \text{ and } t = \tfrac{5}{8}v^2 - \tfrac{5}{8} \times 4^2 = \tfrac{5}{8}v^2 - 10.$$

To find how long it takes her to accelerate to 8 m s^{-1}, put $v = 8$. This gives

$$t = \tfrac{5}{8} \times 64 - 10 = 30.$$

b The original equation now includes an extra term to take account of the resistance, and becomes

$$\frac{80}{v} - 0.8v = 100\frac{dv}{dt}.$$

The left side is $\dfrac{0.8(100-v^2)}{v}$, and $\dfrac{0.8}{100}=\dfrac{1}{125}$, so the equation can be written

$$\frac{dv}{dt}=\frac{100-v^2}{125v}.$$

Inverting this to express $\dfrac{dt}{dv}$ as a function of v gives

$$\frac{dt}{dv}=\frac{125v}{100-v^2},$$

so $t=\displaystyle\int\frac{125v}{100-v^2}\,dv.$

The key to finding this integral is to notice that it can be written as

$$-62.5\times\int\frac{-2v}{100-v^2}\,dv,$$ and that in this form the numerator of the

integrand is the derivative of the denominator with respect to v. This is the special type of integral referred to in P2&3 Section 18.3, and gives the solution

$$t=-62.5\ln(100-v^2)+k.$$

Substituting $t=0$ when $v=4$ gives $0=-62.5\ln 84+k$. Therefore, when $v=8$,

$$t=62.5\,(\ln 84-\ln 36)=52.95\ \dots.$$

The time to accelerate from 4 m s⁻¹ to 8 m s⁻¹ is calculated as

a 30 seconds if resistance is neglected, or

b 53.0 seconds taking resistance into account.

159

It is worth noticing that the answers in Example 9.1.2 could have been obtained directly as definite integrals

a $\displaystyle\int_4^8\frac{5}{4}v\,dv$ and **b** $\displaystyle\int_4^8\frac{125v}{100-v^2}\,dv.$

The reason for not doing this will appear when you come to Example 9.2.2, where algebraic expressions for t in terms of v are needed.

9.2 Displacement–time equations

Often you don't just want to know the velocity of an object at a given time, but also where it is. Once you have the connection between v and t, this can be found by using the fact that $v=\dfrac{dx}{dt}.$

EXAMPLE 9.2.1

In Example 9.1.1, how far does the car travel before it reaches its maximum speed?

The velocity of the car after t seconds is $v = \frac{1}{20}t^{\frac{3}{2}}$ and this equation can be written as

$$\frac{dx}{dt} = \frac{1}{20}t^{\frac{3}{2}}$$

where x metres is the displacement after t seconds. Integrating,

$$x = \int \frac{1}{20}t^{\frac{3}{2}}dt = \frac{1}{50}t^{\frac{5}{2}} + k.$$

Since $x = 0$ when $t = 0$, $k = 0$ and the (t,x) equation is simply $x = \frac{1}{50}t^{\frac{5}{2}}$.

The question asks for the displacement when $v = 10$, that is when $t = 34.19...$ (see Example 9.1.1). Substituting this value for t gives $x = 136.7...$.

The car travels about 137 metres in reaching its maximum speed.

Notice that, if you aren't interested in knowing the (t,x) equation, but only in the total distance travelled, the answer in Example 9.2.1 could be calculated as the definite integral $x = \int_{0}^{34.19...} \frac{1}{20}t^{\frac{3}{2}}dt$. This method is used in the next example.

EXAMPLE 9.2.2

In Example 9.1.2, how far does the cyclist travel in increasing her speed from 4 m s⁻¹ to 8 m s⁻¹?

a if air resistance is neglected,

b making allowance for air resistance of $0.8v$ newtons when her speed is v m s⁻¹?

a If resistance is neglected, t and v are connected by the equation $t = \frac{5}{8}v^2 - 10$.

Before this can be used to find an expression for the displacement, the equation must be rearranged to show v as a function of t. You can easily see that $v^2 = \frac{8}{5}(t+10)$, so

$$v = \sqrt{1.6t + 16}.$$

There is no need to include a ± sign when taking the square root, since all the motion takes place in one direction, which will obviously be chosen to be the positive direction.

The total distance travelled can be found by integrating this with respect to t over the appropriate interval of time. Example 9.1.2(a) shows that this time is from 0 to 30 seconds. The distance is therefore given by

$$x = \int_0^{30} (1.6t + 16)^{\frac{1}{2}} \, dt,$$

which can be evaluated as

$$\left[\frac{1}{1.6} \times \frac{2}{3} (1.6t + 16)^{\frac{3}{2}} \right]_0^{30} = \frac{1}{2.4} \left(64^{\frac{3}{2}} - 16^{\frac{3}{2}} \right) = \frac{1}{2.4} (512 - 64) = 186\tfrac{2}{3},$$

using the rule for integration in P1 Section 16.7.

b Using a similar method to part **a**, the (t,v) equation when resistance is taken into account is $t = 62.5(\ln 84 - \ln (100 - v^2))$, which can be rearranged as

$$-0.016t = \ln \frac{100 - v^2}{84}, \text{ or}$$
$$v = \sqrt{100 - 84e^{-0.016t}}.$$

From Example 9.1.2(b), the range of time is now from 0 to 52.95... seconds, so the distance is given by

$$x = \int_0^{52.95...} \sqrt{100 - 84e^{-0.016t}} \, dt.$$

This is a more difficult integral to evaluate, although it could be found by substitution. However, if you can't see how to do this you can use an approximate method such as the trapezium rule (see P2&3 Chapter 9). With four intervals, this gives the value of the integral as approximately

$$\tfrac{1}{2} \times \tfrac{52.95...}{4} \times (4 + 2(5.659... + 6.708... + 7.450...) + 8),$$

which is about 342.

The distance covered by the cyclist is calculated as approximately **a** 187 metres if resistance is neglected, or **b** 342 metres if resistance is taken into account.

An alternative method for finding the distance is given later, in example 9.3.3. For part **b**, this leads to an expression that can be integrated and so gives an exact solution.

Exercise 9A

1 A car of mass 1200 kg is at rest on a level road. Two people push it, producing a total force given by $(240 - 12t)$ newtons, where t is the time in seconds, until this becomes zero after 20 seconds. How fast is the car then moving, and how far does it move while it is being pushed?

2 A racing car of mass 2000 kg accelerates with a driving force of $480(t - 10)^2$ newtons until it reaches its maximum speed after 10 seconds. Find its maximum speed, and the distance it travels in reaching this speed.

3 A particle of mass 2 kg is at rest at the origin. It is acted on by a force which varies periodically according to the law $F = 8 \sin 2t$ newtons. Find expressions for the velocity and position of the particle after t seconds. Draw sketches of the (t,x) and (t,v) graphs.

4 A car of mass 900 kg is travelling at 24 m s^{-1} when the brakes are suddenly applied. The braking force is given by $(4500 - kt^2)$ newtons, where t is the time in seconds

and k is a constant, and the car comes to a stop in 6 seconds. Find the value of k, and the distance the car travels in coming to a stop.

5   A particle of mass m is at rest at the origin. It is acted on by a force which decreases exponentially according to the law $F = Ke^{-ct}$. Find expressions for its velocity and position after time t. Draw sketches of the velocity–time and displacement–time graphs.

6   A rollerblader of mass 50 kg is moving at 8 m s^{-1} and slows down against a resistance given by v newtons, where v is her speed in m s^{-1}. Find how fast she is moving 20 seconds later, and how far she has gone in that time.

7   An aircraft of mass 4000 kg lands on the deck of an aircraft carrier with a speed of 50 m s^{-1}. It is brought to rest with the help of air brakes and a parachute, which slow it down with a resisting force of $50v^2$ newtons, where v is the speed in m s^{-1}. Find how long it takes for the speed to drop to 10 m s^{-1}, and how far the aircraft travels in this time. Why shouldn't the aircraft rely on this means alone to bring it to rest?

8   A car of mass 1250 kg is taking part in a fuel economy trial. It is conjectured that, if it is travelling at v m s^{-1} with the engine disengaged and without the brakes being applied, the resistance is given by $(5v + v^2)$ newtons. If it is initially moving at 20 m s^{-1}, how long does it take for the speed to drop to 5 m s^{-1}? Find how far the car travels while slowing down to this speed.

9   In first gear a car of mass 900 kg accelerates from rest to 5 m s^{-1} with a constant driving force of 3000 newtons. Assuming air resistance of $30v$ newtons when the car's speed is v m s^{-1}, calculate how long it takes to gain speed, and how far it travels in doing so.

Compare your answers with those obtained by neglecting the resistance.

10   The car in Question 9 accelerates from 5 m s^{-1} to 15 m s^{-1} in second gear, with the engine developing constant power of 15 kW. Neglecting resistance, calculate how long it takes to gain speed, and how far it travels in doing so. Hence find its average speed in this gear. Repeat the calculation if there is air resistance of $30v$ newtons when the speed is v m s^{-1}.

11   An airliner of mass 500 tonnes is powered by engines developing 25 000 kW. Resistance to motion at a speed of v m s^{-1} is $0.8v^2$ newtons. Write as a definite integral the time it takes to reach a speed of 80 m s^{-1} from rest on take-off. Use the trapezium rule to find this time approximately.

12   An underground train of mass 1000 tonnes is accelerated from rest with a driving force of 200 000 newtons. At a speed of v m s^{-1} the air in the tunnel produces a resisting force of $500v^2$ newtons. Find how long the train takes to reach a speed of 10 m s^{-1}.

Write an expression in the form of an integral for the distance it travels in reaching this speed, and use either an exact or an approximate method to estimate this distance.

9.3   Velocity–displacement equations

You are familiar with using displacement–time and velocity–time graphs to show the motion of objects along a line. For example, for the car in Examples 9.1.1 and 9.2.1 these graphs have the forms shown in Fig. 9.1 and Fig. 9.2. The velocity–time graph is

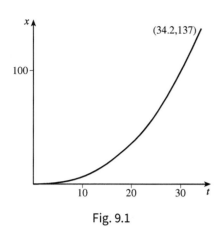

Fig. 9.1

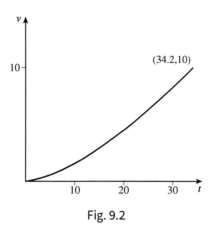

Fig. 9.2

the graph of the gradient of the displacement–time graph, and the acceleration is the gradient of the velocity–time graph.

There is another way of describing the motion, using a velocity–displacement graph. In the example quoted, from the two equations

$$x = \frac{1}{50}t^{\frac{5}{2}} \quad \text{and} \quad v = \frac{1}{20}t^{\frac{3}{2}}$$

found in Examples 9.1.1 and 9.2.1, you can eliminate the variable t.

From $x = \frac{1}{50}t^{\frac{5}{2}}$, you can deduce $t = (50x)^{0.4}$. So

$$t^{\frac{3}{2}} = \left((50x)^{0.4}\right)^{\frac{3}{2}} = (50x)^{0.4 \times \frac{3}{2}} = (50x)^{0.6}$$

$$= 10.456...x^{0.6}$$

which gives

$$v = \frac{1}{20}t^{\frac{3}{2}} = 0.5228...x^{0.6}.$$

This graph is shown, for values of v from 0 to 10, in Fig. 9.3.

How can you find the acceleration from a velocity–displacement graph? To answer this, notice that the (t,x) and (t,v) equations above can be regarded as parametric equations for the (x,v) graph, with t as the parameter. So the gradient of the (x,v) graph can be found by the rule proved in P2&3 Section 10.3, as

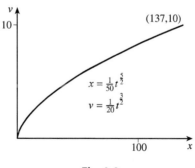

Fig. 9.3

$$\frac{dv}{dx} = \frac{dv}{dt} \Big/ \frac{dx}{dt}.$$

Since $\dfrac{dv}{dt} = a$ and $\dfrac{dx}{dt} = v$, this is $\dfrac{dv}{dx} = \dfrac{a}{v}$,

so $a = v\dfrac{dv}{dx}.$

This is an important result, which you should remember alongside $v = \dfrac{dx}{dt}$ and $a = \dfrac{dv}{dt} = \dfrac{d^2x}{dt^2}$ which were established in M1 Chapter 11.

> **For an object moving in a straight line, if x denotes the displacement from a fixed point O of the line, v denotes the velocity and a the acceleration, then**
> $$a = v\frac{dv}{dx}.$$

EXAMPLE 9.3.1

A particle moves along a straight line in such a way that the velocity when it has travelled a distance x is given by $v = \dfrac{1}{p + qx}$, where p and q are constants.

Find expressions for the acceleration   **a** in terms of x,   **b** in terms of v.

a $\quad \dfrac{dv}{dx} = -\dfrac{q}{(p+qx)^2}, \quad$ so $\quad a = v\dfrac{dv}{dx} = -\dfrac{q}{(p+qx)^3}.$

b $\quad \dfrac{1}{(p+qx)^3} = v^3, \quad$ so $\quad a = -qv^3.$

If you have an expression for the acceleration in terms of either x or v (or both), you can use the form $v\dfrac{dv}{dx}$ to make a differential equation connecting x and v, without bringing in time.

EXAMPLE 9.3.2

A projectile is launched vertically upwards from the surface of the moon with initial speed u. The radius of the moon is R. When the projectile is at a height x above the surface, the gravitational attraction produces an acceleration $\dfrac{C}{(R+x)^2}$ towards the centre of the moon, where C is a positive constant. Find an expression for the speed of the projectile when it is at height x. How large must u be for the projectile never to return to the surface?

Taking the upward direction to be positive,

$$v\frac{dv}{dx} = -\frac{C}{(R+x)^2}.$$

This is a differential equation of the separable variables type (see P2&3 Section 19.5), but it is already arranged so that it can be directly integrated with respect to x.

Since $v\dfrac{\mathrm{d}v}{\mathrm{d}x}=\dfrac{\mathrm{d}}{\mathrm{d}x}\left(\tfrac{1}{2}v^2\right)$ by the chain rule,

$$\frac{1}{2}v^2=\int-\frac{C}{(R+x)^2}\,\mathrm{d}x=\frac{C}{R+x}+k.$$

It is given that $v=u$ when $x=0$, so

$$\frac{1}{2}u^2=\frac{C}{R}+k.$$

Therefore $\dfrac{1}{2}v^2-\dfrac{1}{2}u^2=\dfrac{C}{R+x}-\dfrac{C}{R}$, giving $v=\pm\sqrt{\dfrac{2C}{R+x}-\dfrac{2C}{R}+u^2}$.

There are now two possibilities according to the value of u. The term $\dfrac{2C}{R+x}$ gets smaller as x increases, but it always stays positive. So if $u^2\geq\dfrac{2C}{R}$, v never becomes zero; and since it has the positive value u to start with, it is always positive. This means that if $u\geq\sqrt{\dfrac{2C}{R}}$ the projectile goes on moving away from the moon indefinitely, and never returns to the surface.

If $u<\sqrt{\dfrac{2C}{R}}$, v becomes zero when $\dfrac{2C}{R+x}-\dfrac{2C}{R}+u^2=0$. This equation can be solved to give $x=\dfrac{u^2R}{\dfrac{2C}{R}-u^2}$. Because the expression inside the square root sign can't be negative, x can never be greater than this. So what happens is that the velocity changes sign from $+$ to $-$, which means that the projectile falls back to the surface of the moon. Notice that, since x is the only variable in the expression for v, the value of $|v|$ is the same at any given height whether the projectile is going up or coming down. When it hits the surface, it is again moving with speed u.

For the problem of the cyclist in Example 9.1.2, you can use the form $v=\dfrac{\mathrm{d}v}{\mathrm{d}x}$ for the acceleration to find the distance travelled directly, without first finding the time. You should compare the answers in the next example with those found in Example 9.2.2.

EXAMPLE 9.3.3

In Example 9.1.2, how far does the cyclist travel in increasing her speed from 4 m s^{-1} to 8 m s^{-1} in the two cases **a** and **b**?

The differential equations are formed in just the same way as in Example 9.1.2, but writing the acceleration as $v=\dfrac{\mathrm{d}v}{\mathrm{d}x}$ rather than $\dfrac{\mathrm{d}v}{\mathrm{d}t}$.

a If resistance is neglected,

$$\frac{80}{v} = 100v\frac{dv}{dx}.$$

Since v appears in this equation but x doesn't, you want the derivative to appear as $\frac{dx}{dv}$ rather than $\frac{dv}{dx}$. So using $\frac{dx}{dv} = 1 \bigg/ \frac{dv}{dx}$, the differential equation can be arranged as $\frac{dx}{dv} = \frac{5}{4}v^2$.

Only the total distance is asked for, so this can be calculated as the definite integral

$$\int_4^8 \frac{5}{4}v^2 dv = \left[\frac{5}{12}v^3\right]_4^8 = \frac{5}{12}(512 - 64) = 186\frac{2}{3}.$$

b Taking resistance into account,

$$\frac{80}{v} - 0.8v = 100v\frac{dv}{dx}.$$

The left side is $\dfrac{0.8\left(100 - v^2\right)}{v}$, so the differential equation can be arranged as

$$\frac{dx}{dv} = \frac{125v^2}{100 - v^2}.$$

The distance is then calculated as

$$\int_4^8 \frac{125v^2}{100 - v^2} dv.$$

The key to finding this integral is to write the v^2 in the numerator as $100 - (100 - v^2)$,

so that the integrand then becomes $\dfrac{12500}{100 - v^2} - 125$. The first term can then be split

into partial fractions as $\dfrac{625}{10 + v} + \dfrac{625}{10 - v}$. The integral then becomes

$$\int_4^8 \left(\frac{625}{10 + v} + \frac{625}{10 - v} - 125\right) dv = [625\ln(10 + v) - 625\ln(10 - v) - 125v]_4^8$$

$$= 625(\ln 18 - \ln 14 - \ln 2 + \ln 6) - 125(8 - 4)$$

$$= 625\ln\tfrac{27}{7} - 500 = 343.7...$$

The distance covered by the cyclist is calculated as 187 metres if resistance is neglected, or as 344 metres if resistance is taken into account.

Notice that the answer to part **b** of Example 9.2.2, 342 metres, is very close to the value given by this exact method. This suggests that using the trapezium rule was quite acceptable, bearing in mind that the model on which the calculation is based is itself only approximate.

9.4 Reintroducing time

Sometimes you want to reverse the process described at the beginning of Section 9.3, and find the (t,x) or the (t,v) relation from an equation connecting velocity and displacement. This can be done by writing v as $\dfrac{\mathrm{d}x}{\mathrm{d}t}$ and solving a differential equation involving x and t.

EXAMPLE 9.4.1

A car is travelling at 10 m s^{-1} when the driver applies the brakes and brings the car to rest in a distance of 20 metres. The velocity–displacement relationship is modelled by a straight line graph. Find an expression for the distance the car has travelled t seconds after the brakes are applied.

Fig. 9.4 shows the velocity–displacement graph, which has equation $2v + x = 20$

Writing v as $\dfrac{\mathrm{d}x}{\mathrm{d}t}$, this gives the differential equation

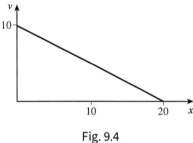

Fig. 9.4

$$\frac{\mathrm{d}x}{\mathrm{d}t} = \frac{1}{2}(20 - x).$$

Since the right side is a function of x rather than t, use $\dfrac{\mathrm{d}t}{\mathrm{d}x} = 1 \Big/ \dfrac{\mathrm{d}x}{\mathrm{d}t}$ to rewrite the differential equation as

$$\frac{\mathrm{d}t}{\mathrm{d}x} = \frac{2}{20 - x}.$$

This has solution

$$t = -2\ln(20 - x) + k.$$

Since t is measured from the instant when the brakes are applied, $t = 0$ when $x = 0$.

So $0 = -2\ln 20 + k$, and

$$t = -2\ln(20 - x) + 2\ln 20,$$

or more conveniently

$$t = -2\ln\frac{20 - x}{20} = -2\ln\left(1 - \frac{1}{20}x\right)$$

The question asks for x in terms of t, so the equation must now be rearranged in the form $x = \ldots$. This can be done by writing the equation in exponential form as

$$e^{-\frac{1}{2}t} = 1 - \frac{1}{20}x \text{ , giving } x = 20\left(1 - e^{-\frac{1}{2}t}\right).$$

It is a good idea to check the answer by differentiating this equation with respect to t, to get

$$v = \frac{dx}{dt} = 10e^{-\frac{1}{2}t}, \text{ so } 2v + x = 20e^{-\frac{1}{2}t} + 20\left(1 - e^{-\frac{1}{2}t}\right) = 20,$$

which brings you back to the original (x, v) equation.

The (t,x) and (t,v) equations show that, with the suggested model, x never reaches the value 20 in finite time, and v never becomes zero. So this is not a plausible model for the complete braking operation, although it might be valid over most of the distance.

9.5 Work done by a force

The work–energy equation

$$Fs = \tfrac{1}{2}mv^2 - \tfrac{1}{2}mu^2 \qquad \text{(see M1 Section 8.1)}$$

was obtained by combining the equation $F = ma$ with the constant acceleration formula $v^2 = u^2 + 2as$. So in this form it only applies to situations in which the force is constant. Using the expression $v\dfrac{dv}{dx}$ for the acceleration, the equation can be generalised to cover situations in which the force is variable.

Integrating Newton's second law in the form $F = mv\dfrac{dv}{dx}$ leads to

$$\int_{x_1}^{x_2} F \, dx = \int_{x_1}^{x_2} mv\frac{dv}{dx} \, dx = \int_{v_1}^{v_2} mv \, dv = \left[\tfrac{1}{2}mv^2\right]_{v_1}^{v_2} = \tfrac{1}{2}mv_2{}^2 - \tfrac{1}{2}mv_1{}^2.$$

The last expression is the gain in the kinetic energy of the object as it moves from $x = x_1$ to $x = x_2$ so $\int_{x_1}^{x_2} F \, dx$ is the work done by the force.

If an object modelled as a particle moving in a straight line changes its position from x_1 to x_2 under the action of a force F, the work done by the force is $\int_{x_1}^{x_2} F \, dx$.

This proves the statement in Section 5.3, that the work done by a force is represented by the area under the force–displacement graph.

Notice also that, if W denotes the work done up to the position x, you can go from the equation

$$W = \int_{x_1}^{x} F \, dx$$

to the derivative

$$\frac{\mathrm{d}W}{\mathrm{d}x} = F,$$

so that $\dfrac{\mathrm{d}W}{\mathrm{d}t} = \dfrac{\mathrm{d}W}{\mathrm{d}x} \times \dfrac{\mathrm{d}x}{\mathrm{d}t} = Fv.$

This justifies the statement in M1 Section 8.5 that the power developed by an engine producing a driving force F, defined as the rate at which the engine does work, is equal to Fv. This result has already been used in this chapter in Example 9.1.2.

Exercise 9B

1 For the following equations and intervals, find equations connecting v with x. Draw sketches of the (t,x), (t,v) and (x,v) graphs.

   a $x = 4t^2 + t$ for $0 \leq t \leq 2$

   b $x = 3\sin 2t$ for $0 \leq t \leq 2\pi$

   c $x = 5\mathrm{e}^{-2t}$ for $0 \leq t \leq \ln 2$

   d $x = \dfrac{1}{t+1}$ for $0 \leq t \leq 4$

2 A particle is at the origin at time $t = 0$, and its velocity is given by the following equations. Find equations connecting v with x, and find expressions for the acceleration

   i in terms of x,   **ii** in terms of v.

   a $v = \sec^2 t$ for $0 \leq t \leq \frac{1}{2}\pi$

   b $v = \mathrm{e}^{\frac{1}{4}t}$ for $t \geq 0$

   c $v = \sin\frac{1}{2}t$ for $t \geq 0$

   d $v = \mathrm{e}^t + \mathrm{e}^{-t}$ for $t \geq 0$

3 For the rollerblader in Exercise 9A Question 6, find an equation connecting v with x.

4 For the aircraft in Exercise 9A Question 7, use a direct method to find the distance travelled as the speed drops from 50 m s^{-1} to 10 m s^{-1}.

5 For the car in Exercise 9A Question 8, use a direct method to find the distance travelled while the speed drops from 20 m s^{-1} to 5 m s^{-1}.

6 For the train in Exercise 9A Question 12, use a direct method to find the distance travelled in reaching a speed of 10 m s^{-1} from rest.

7 A motorcycle with its rider has mass 300 kg. The power of the engine is 5 kW, and the air resistance is given by $0.5v^2$ newtons when the speed is v m s^{-1}. Find how far it travels in increasing its speed from 5 m s^{-1} to 15 m s^{-1}.

8 For the airliner in Exercise 9A Question 11, find the distance travelled in reaching the take-off speed of 80 m s^{-1} from rest.

9 On joining a motorway a car of mass 1800 kg accelerates from 10 m s^{-1} to 30 m s^{-1}. The engine produces a constant driving force of 4000 newtons, and the resistance to motion at a speed of v m s^{-1} is $0.9v^2$ newtons. Find how far the car travels while accelerating.

10 A car of mass 4000 kg, trying to beat the land speed record, reaches a speed of 300 m s^{-1}. In order to slow down from this speed, the driver first deploys a small parachute, which produces a resistance of $2v^2$ newtons at a speed of v m s^{-1}. When the speed has dropped to 100 m s^{-1}, mechanical brakes are applied which increase the retarding force by a further 8000 newtons. What is the total distance required to bring the car to a complete stop?

11   The (x,v) equation for a particle is $v = 10 - 0.1\,x^2$ for $0 < x < 10$. Find the (t,x) equation, given that $x = 0$ when $t = 0$. Check your answer by finding an expression for v both as $\dfrac{dx}{dt}$ and as $10 - 0.1x^2$.

12   The (x,v) equation for a particle is $v = \sqrt{p + qx}$ where p and q are constants. Show that the acceleration is constant. Find the (t,x) equation, given that $x = 0$ when $t = 0$, and express it in as simple a form as possible.

   By making suitable substitutions for p and q, identify your equations with the standard equations for motion with constant acceleration.

13   If $x = c \sin (nt + \varepsilon)$, where c, n and ε are constants, obtain the (x,v) equation and express the acceleration in terms of x. Draw sketches of the (t,x), (t,v) and (x,v) graphs.

14   A piston is used to compress the air in a cylinder 0.2 metres long. When the piston has moved x metres, the force opposing it is $\dfrac{10}{0.2 - x}$ newtons. Calculate the work done in moving the piston a distance of 0.18 metres.

15   A girl sets a gyroscope spinning by means of a thread 80 cm long. When she has pulled it through x cm, she is exerting a force of $\left(20 + \frac{1}{4}x\right)$ newtons. Find the kinetic energy of the gyroscope when the whole thread has been pulled through.

9.6   Vertical motion with air resistance

A particular application of the methods described in Sections 9.1 to 9.4 is to objects moving vertically under gravity and opposed by air resistance depending on the speed. The basic ideas were explained in M1 Section 6.4, but at that stage it was only possible to give a numerical and graphical treatment. Now you can find equations which fit each of the most commonly used models. These models are that the resistance is proportional to the square of the speed, and that it is proportional to the speed.

In this section these models are applied to the problem discussed in M1 Example 6.4.2.

EXAMPLE 9.6.1

A cannonball is projected vertically upwards from a mortar with an initial speed of 40 m s^{-1}. The mortar is situated at the edge of a cliff 100 metres above the sea. On the way down, the cannonball just misses the cliff. In vertical fall the cannonball would have a terminal speed of 50 m s^{-1}. Assuming that air resistance is proportional to the square of the speed, find how high the cannonball rises and how long it is in the air before falling into the sea.

   Suppose that the mass of the cannonball is M kg, and that the air resistance at speed v m s^{-1} is kv^2 newtons. The constant k can be found from the terminal speed; at this speed the weight of the cannonball is equal to the resistance, so

$$10M = k \times 50^2, \quad \text{which gives} \quad k = 0.004M.$$

The resistance $0.004Mv^2$ is always positive or zero. So because it acts downwards when the cannonball is going up, and upwards when it is going down, the problem must be split into two stages: the motion upwards to the highest point, and the drop from the highest point into the sea.

Upward stage While the cannonball is on the way up, it is opposed by both the weight and the resistance. Newton's second law then gives

$$-10M - 0.004Mv^2 = Ma,$$

where a m s^{-2} is the acceleration and upwards is taken as the positive direction.

Therefore $a = -0.004(2500 + v^2)$.

To find the height, you want an equation connecting velocity and displacement,

so write a as $v\dfrac{dv}{dx}$. Then

$$v\frac{dv}{dx} = -0.004\left(2500 + v^2\right), \text{ which can be rearranged as } \frac{dx}{dv} = -\frac{250v}{2500 + v^2}.$$

Initially the velocity is 40, and at maximum height this has dropped to zero. So the total height is given by the definite integral

$$\int_{40}^{0} -\frac{250v}{2500 + v^2}\,dv.$$

The minus sign in the integrand can be used to reverse the order of the limits of integration, and the integral can then be calculated as

$$\int_{0}^{40} \frac{250v}{2500 + v^2}\,dv = \left[125\ln(2500 + v^2)\right]_0^{40} = 125\left(\ln 4100 - \ln 2500\right)$$

$$= 125\ln\left(\tfrac{4100}{2500}\right) = 125\ln 1.64 = 61.8....$$

You also need to find how long the cannonball takes to reach its highest point. To do this, write a as $\dfrac{dv}{dt}$, so

$$\frac{dv}{dt} = -0.004(2500 + v^2), \text{ which can be rearranged as } \frac{dt}{dv} = -\frac{250}{2500 + v^2}.$$

The time up to the highest point is therefore given by

$$\int_{40}^{0} -\frac{250}{2500 + v^2}\,dv, \text{ which is } \int_{0}^{40} \frac{250}{2500 + v^2}\,dv.$$

You can evaluate this integral by using the substitution $v = 50\tan\theta$

(see P2&3 Chapter 18). Then $\dfrac{dv}{d\theta} = 50\sec^2\theta$, so

$$\int_{0}^{40} \frac{250}{2500 + v^2}\,dv = \int_{0}^{\tan^{-1}0.8} \frac{250}{2500 + 2500\tan^2\theta} \times 50\sec^2\theta\,d\theta$$

$$= \int_{0}^{\tan^{-1}0.8} \frac{250}{2500\sec^2\theta} \times 50\sec^2\theta\,d\theta$$

$$= \int_{0}^{\tan^{-1}0.8} 5\,d\theta = 5\tan^{-1}0.8 = 3.373....$$

Downward stage When the cannonball starts to come down, the direction of the resistance is reversed, and new equations are needed. It is also best to choose the positive direction to be downwards, and to measure x and t from the instant when the cannonball is at its highest point. The equations then become

$$\frac{\mathrm{d}x}{\mathrm{d}v} = \frac{250v}{2500 - v^2} \quad \text{and} \quad \frac{\mathrm{d}t}{\mathrm{d}v} = \frac{250}{2500 - v^2},$$

with $x = 0$ and $v = 0$ when $t = 0$.

You now know that the cannonball has to fall a distance $(61.8\ldots + 100)$ metres into the sea, and you want to know the time this takes. That is, you want the value of t when $x = 161.8.\ldots$ So it seems best to begin by using the equation which involves t. Splitting the right side into partial fractions gives

$$\frac{\mathrm{d}t}{\mathrm{d}v} = \frac{5}{2}\left(\frac{1}{50 + v} + \frac{1}{50 - v}\right),$$

which can be integrated as

$$t = \frac{5}{2}(\ln(50 + v) - \ln(50 - v)) + k = \frac{5}{2}\ln\frac{50 + v}{50 - v} + k$$

Since $v = 0$ when $t = 0$, $0 = \frac{5}{2}\ln 1 + k = 0 + k.$ So $k = 0$, and

$$t = \frac{5}{2}\ln\frac{50 + v}{50 - v}.$$

This is halfway to the answer, but to find the (t, x) equation you have to write v as $\dfrac{\mathrm{d}x}{\mathrm{d}t}$ and integrate a second time. First, put the equation into exponential form as

$$\frac{50 + v}{50 - v} = e^{0.4t}, \text{ and then rearrange this to get } \frac{\mathrm{d}x}{\mathrm{d}t} = v = 50\frac{e^{0.4t} - 1}{e^{0.4t} + 1}.$$

At this point you may hit a snag. Further progress depends on being able to integrate the right side of this equation, which requires a small trick. In case you can't spot this, it is useful to have an alternative method to turn to. So here are two ways of finishing off the problem.

Method 1 The trick you need is to divide top and bottom of the fraction on the right side by $e^{0.2t}$, so the equation becomes

$$\frac{\mathrm{d}x}{\mathrm{d}t} = 50\frac{e^{0.2t} - e^{-0.2t}}{e^{0.2t} + e^{-0.2t}}.$$

You now have a fraction in which the numerator is almost (apart from a constant factor) the derivative of the denominator, and it isn't hard to see that

$$x = \frac{50}{0.2}\ln\left(e^{0.2t} + e^{-0.2t}\right) + k$$

Since $x = 0$ when $t = 0$, $0 = 250 \ln 2 + k$, so $k = -250 \ln 2$.

Therefore $x = 250\ln\left(\tfrac{1}{2}\left(e^{0.2t} + e^{-0.2t}\right)\right).$

All that remains is to solve this equation for t when $x = 161.8\ldots$. So put the equation into exponential form as

$$e^{0.2t} + e^{-0.2t} = 2e^{\frac{161.8\ldots}{250}} = 3.82..$$

and notice that if you multiply this by $e^{0.2t}$ you get a quadratic equation with $e^{0.2t}$ as the unknown:

$$\left(e^{0.2t}\right)^2 - 3.82\ldots e^{0.2t} + 1 = 0.$$

The solution of this is $e^{0.2t} = 3.53\ldots$ or $0.28\ldots$, giving

$$t = \frac{\ln 3.53\ldots}{0.2} = 6.318\ldots \quad \text{or} \quad t = \frac{\ln 0.28\ldots}{0.2} = -6.318\ldots.$$

Since t must be positive, the required value of t is $6.318\ldots$.

Method 2 Instead of trying to connect t and x directly, this method links the two variables by finding the speed when the cannonball enters the sea. To do this, go back to the (x, v) equation

$$\frac{dx}{dv} = \frac{250v}{2500 - v^2},$$

which can be integrated directly as

$$x = -125 \ln(2500 - v^2) + k.$$

Since $v = 0$ when $x = 0, 0 = -125 \ln 2500 + k$. So $k = 125 \ln 2500$, and

$$x = -125 \ln(2500 - v^2) + 125 \ln 2500 = -125 \ln\left(1 - \tfrac{1}{2500} v^2\right).$$

This can be rearranged to give

$$1 - \frac{1}{2500} v^2 = e^{-0.008x}, \quad \text{so} \quad v = 50\sqrt{1 - e^{-0.008x}}.$$

You can now calculate that, when $x = 161.8\ldots$, $v = 42.60\ldots$. This value of v can then be substituted in the equation for t found above, to give

$$t = \frac{5}{2} \ln \frac{50 + v}{50 - v} = \frac{5}{2} \ln \frac{50 + 42.60\cdots}{50 - 42.60\cdots} = 6.318\cdots$$

Whichever method you use, this value of t now has to be added to the time found earlier for the upward stage to give a total time of $(3.373\ldots + 6.318\ldots)$ seconds, which is $9.691\ldots$ seconds.

So, with this model, the cannonball rises to a height of 61.8 metres and then falls into the sea after being in the air for 9.7 seconds.

EXAMPLE 9.6.2

Rework Example 9.6.1, assuming that air resistance is proportional to the speed.

If the air resistance at speed v m s^{-1} is kv newtons, equating the weight to the resistance at terminal speed gives

$$10M = k \times 50, \quad \text{so} \quad k = 0.2M.$$

For the upward stage, Newton's second law then gives

$$-10M - 0.2Mv = Ma, \quad \text{so} \quad a = -10 - 0.2v.$$

What about the downward stage? As in Example 9.6.1, the direction of the resistance is reversed when the cannonball starts to fall. But, if a, v, x and t keep the same meaning as they had for the upward stage, the velocity v changes sign from positive to negative at the same instant. So the equation for the acceleration for the downward stage becomes

$$a = -10 + 0.2|v|, \quad \text{where } v = |v|;$$

that is,

$$a = -10 - 0.2v,$$

the same equation as for the upward stage.

What this shows is that, if the air resistance is proportional to the speed, you can use the same equation for the downward stage as for the upward stage, with all the variables having the same meaning for both stages.

It is now very easy to complete the solution. The differential equation

$$\frac{dv}{dt} = -0.2(50 + v) \text{ can be rearranged as } \frac{dt}{dv} = -5 \times \frac{1}{50 + v}.$$

Integrating, $t = -5\ln(50 + v) + k$.

You know that $v = 40$ when $t = 0$, so $0 = -5\ln 90 + k$. Therefore

$$t = -5\ln\frac{50 + v}{90}.$$

In exponential form, this is

$$v = 90e^{-0.2t} - 50.$$

Writing v as $\dfrac{dx}{dt}$ a second integration gives

$$x = -450e^{-0.2t} - 50t + c.$$

Since $x = 0$ when $t = 0$, $c = 450$, so

$$x = -450e^{-0.2t} + 450 - 50t.$$

These equations can now be used to find the greatest height and the time that the cannonball is in the air. The greatest height is reached when $v = 0$, that is when

$$e^{-0.2t} = \frac{5}{9}, \quad \text{or} \quad t = \frac{\ln 1.8}{0.2}.$$

Then $x = -450 \times \frac{5}{9} + 450 - 250\ln 1.8 = 53.05\ldots$

The cannonball enters the sea when $x = -100$, which leads to the equation

$$450e^{-0.2t} + 50t = 550.$$

> This can't be solved exactly, but you can use an approximate numerical method (see P2&3 Chapter 8) to find that the solution is 9.70... .
>
> So, with this model, the cannonball rises to a height of 53.1 metres and falls into the sea after being in the air for 9.7 seconds.

It is interesting to notice that, although the values found for the maximum height are very different for the two models and for the calculation which ignores the resistance entirely (61.8 metres and 53.1 metres, compared with 80 metres), those for the total time are very close (9.7 seconds in both these examples, compared with 10 seconds if resistance is ignored).

Exercise 9C

1 A rock of mass 10 kg falls over a cliff and drops vertically on to a field 200 metres below. The air resistance is given by $0.04\,v^2$ newtons when the speed is v m s^{-1}. Find

   a   the terminal speed for a fall of indefinite distance,

   b   the speed with which the rock hits the field,

   c   the time that the rock takes to fall.

2 A bullet of mass 50 grams is fired vertically into the air with a speed of 600 m s^{-1}. At a speed of v m s^{-1} it experiences air resistance of kv^2 newtons, where $k = 5 \times 10^{-5}$. Find

   a   how high the bullet rises,

   b   the speed of the bullet at half this height,

   c   the time taken by the bullet to reach its maximum height.

3 A stone of mass 100 grams will fall with terminal speed 40 m s^{-1}. A boy catapults the stone vertically upwards with a speed of 25 m s^{-1}. Assuming that the resistance to motion is proportional to the speed, find

   a   how high the stone rises,

   b   how fast the stone is moving just before it hits the ground,

   c   how long it takes for the stone to return to ground level.

4 Repeat Question 3 if the air resistance is proportional to the square of the speed.

Miscellaneous exercise 9

1 A particle P of mass 0.5 kg moves along the x-axis on a horizontal surface. When the displacement of P from the origin O is x m the velocity of P is v m s^{-1} in the positive x-direction. Two horizontal forces act on P; one force has magnitude $(1 + 0.3x^2)$ N and acts in the positive x-direction, and the other force has magnitude $8e^{-x}$ N and acts in the negative x-direction.

   i   Show that $v\dfrac{\mathrm{d}v}{\mathrm{d}x} = 2 + 0.6x^2 - 16e^{-x}$.

   ii   The velocity of P as it passes through O is 6 m s^{-1}. Find the velocity of P when $x = 3$.

(Cambridge International AS and A Level Mathematics 9709/05 Paper 5 Q3 November 2008)

175

2 A particle P starts from rest at a point O and travels in a straight line. The acceleration of P is $(15 - 6x)$ m s^{-2}, where x m is the displacement of P from O.

   i   Find the value of x for which P reaches its maximum velocity, and calculate this maximum velocity.

   ii Calculate the acceleration of P when it is at instantaneous rest and $x > 0$.

   (Cambridge International AS and A Level Mathematics 9709/51 Paper 5
   Q4 June 2011)

3 A particle P of mass 0.8 kg moves along the x-axis on a horizontal surface. When the displacement of P from the origin O is x m the velocity of P is v m s^{-1} in the positive x-direction. Two horizontal forces act on P. One force has magnitude $4e^{-x}$ N and acts in the positive x-direction. The other force has magnitude $2.4x^2$ N and acts in the negative x-direction.

   i   Show that $v\dfrac{dv}{dx} = 5e^{-x} - 3x^2$.

   ii The velocity of P as it passes through O is 6 m s^{-1}. Find the velocity of P when $x = 2$.

   (Cambridge International AS and A Level Mathematics 9709/51 Paper 5
   Q3 November 2013)

4 A particle of mass 0.2 kg is projected vertically downwards with initial speed 4 m s^{-1}. A resisting force of magnitude $0.09v$ N acts vertically upwards on the particle during its descent, where v m s^{-1} is the downwards velocity of the particle at time t s after being set in motion.

   i   Show that the acceleration of the particle is $(10 - 0.45v)$ m s^{-2}.

   ii Find v when $t = 1.5$.

   (Cambridge International AS and A Level Mathematics 9709/51 Paper 5
   Q4 June 2013)

5 A ball of mass 0.05 kg is released from rest at a height h m above the ground. At time t s after its release, the downward velocity of the ball is v m s^{-1}. Air resistance opposes the motion of the ball with a force of magnitude $0.01v$ N.

   i   Show that $\dfrac{dv}{dt} = 10 - 0.2v$. Hence find v in terms of t.

   ii Given that the ball reaches the ground when $t = 2$, calculate h.

   (Cambridge International AS and A Level Mathematics 9709/51 Paper 5
   Q5 November 2011)

6 A particle P of mass 0.4 kg is released from rest at the top of a smooth plane inclined at $30°$ to the horizontal. The motion of P down the slope is opposed by a force of magnitude $0.6x$ N, where x m is the distance P has travelled down the slope. P comes to rest before reaching the foot of the slope. Calculate

   i   the greatest speed of P during its motion,

   ii the distance travelled by P during its motion.

   (Cambridge International AS and A Level Mathematics 9709/51 Paper 5
   Q5 June 2012)

7 A particle of mass 0.25 kg moves in a straight line on a smooth horizontal surface. A variable resisting force acts on the particle. At time t s the displacement of the particle from a point on the line is x m, and its velocity is $(8 - 2x)$ ms^{-1}. It is given that $x = 0$ when $t = 0$.

   i   Find the acceleration of the particle in terms of x, and hence find the magnitude of the resisting force when $x = 1$.

  ii   Find an expression for x in terms of t.

 iii   Show that the particle is always less than 4 m from its initial position.

(Cambridge AS & A Level Mathematics 9709/05 Paper 5
Q7 November 2005)

8 A cyclist starts from rest at a point O and travels along a straight path. At time t s after starting, the displacement of the cyclist from O is x m, and the acceleration of the cyclist is a ms^{-2}, where $a = 0.6x^{0.2}$.

   i   Find an expression for the velocity v ms^{-1} of the cyclist in terms of x.

  ii   Show that $t = 2.5x^{0.4}$.

 iii   Find the distance travelled by the cyclist in the first 10 s of the journey.

(Cambridge AS & A Level Mathematics 9709/05 Paper 5
Q7 November 2006)

9 A particle P of mass 0.5 kg moves on a horizontal surface along the straight line OA, in the direction from O to A. The coefficient of friction between P and the surface is 0.08. Air resistance of magnitude $0.2v$ N opposes the motion, where v m s^{-1} is the speed of P at time t s. The particle passes through O with speed 4 m s^{-1} when $t = 0$.

   i   Show that $2.5\dfrac{\mathrm{d}v}{\mathrm{d}t} = -(v + 2)$ and hence find the value of t when $v = 0$.

  ii   Show that $\dfrac{\mathrm{d}x}{\mathrm{d}t} = 6e^{-0.4t} - 2$, where x m is the displacement of P from O at time t s,

and hence find the distance OP when $v = 0$.

(Cambridge International AS and A Level Mathematics 9709/05 Paper 5
Q7 June 2008)

10 A particle P of mass 0.25 kg moves in a straight line on a smooth horizontal surface. P starts at the point O with speed 10 m s^{-1} and moves towards a fixed point A on the line. At time t s the displacement of P from O is x m and the velocity of P is v m s^{-1}. A resistive force of magnitude $(5 - x)$ N acts on P in the direction towards O.

   i   Form a differential equation in v and x. By solving this differential equation, show that $v = 10 - 2x$.

  ii   Find x in terms of t, and hence show that the particle is always less than 5 m from O.

(Cambridge International AS and A Level Mathematics 9709/51 Paper 5
Q7 June 2010)

Chapter 10
Strategies for solving problems

This chapter is about choosing the method to use when you have a problem to solve. When you have completed it, you should

- be able to judge which methods are appropriate in a particular problem, and the relative advantages of using various methods.

You already know that there may be more than one way of solving some problems in mechanics. The aim of this chapter is to help you decide how to tackle any particular problem. There is no essentially new mechanics in this chapter, but the examples have been chosen to bring out the advantages and disadvantages of different methods.

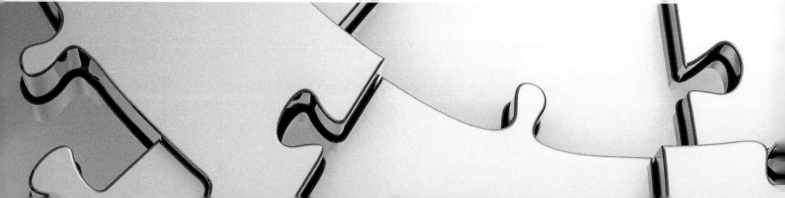

10.1 Equilibrium problems

When you have a problem involving an object in equilibrium, the first decision is whether to use an algebraic method (equations of moments or resolving or both) or a geometrical method. Sometimes this is just a matter of personal preference, but some problems are more approachable by one method or the other.

Geometrical methods are generally only useful when there are just two or three forces. If two forces are in equilibrium, they have equal magnitude and act along the same line. If three forces are in equilibrium, their lines of action are either concurrent or parallel, and they may be represented in a vector triangle of forces. However, if there are more than three forces it may be worth combining pairs of forces to reduce the total number; for example, replacing a normal contact force and a frictional force by a total contact force, or replacing two weights by a single weight at the centre of mass.

If you are using an algebraic method, you have a wide choice of possible directions to resolve in, and of points to take moments about. A good choice may lead to an equation involving only one or two of the unknowns, which can improve the efficiency of the solution.

If an object is supported by a hinge, you won't know either the magnitude or the direction of the force it exerts. Usually, if you are using a geometrical method, it is best to treat it as a single force; but with an algebraic method it is often better to replace the single force by its components in two perpendicular directions.

These points are illustrated in the examples that follow.

EXAMPLE 10.1.1

A uniform ladder of weight W and length l rests at 20° to the vertical, with one end on horizontal ground and the other end against a vertical wall. A tie-bar has one end fixed at the corner C where the wall meets the ground; the other end is fixed to the ladder so that the tie-bar is perpendicular to the ladder. If there are no frictional forces, find the tension in the bar.

There are four forces on the ladder (Fig. 10.1), and there is no good reason to combine any pair of them, so a geometrical method isn't suitable.

Only the tension T and the weight W are of interest, so it is a good idea to write an equation which doesn't bring in either of the normal contact forces. You can do this by taking moments about the point P where the lines of action of these two forces meet.

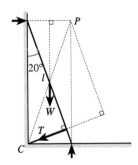

Fig. 10.1

The ladder is described as 'uniform', which means that its centre of mass is at its mid-point. The distance from P to the line of action of W is therefore $\frac{1}{2}l\sin 20°$.

The wall, the ground and the lines of the two normal forces form a rectangle whose diagonals are CP and the ladder. Therefore CP has length l and makes an angle of 20° with the vertical. So the angle between CP and the tie-bar is

$90° - 2 \times 20° = 50°$, and the perpendicular distance from P to the line of action of T is $l \sin 50°$.

$$M(P) \; W \times \tfrac{1}{2} l \sin 20° = T \times l \sin 50°,$$

so $\quad T = \dfrac{W \sin 20°}{2 \sin 50°} = 0.223W$, correct to 3 significant figures

---

EXAMPLE 10.1.2

A uniform ladder AB is set up against a wall. The coefficient of friction μ is the same for both contacts, where $\mu < 1$. Prove that, if equilibrium is limiting, the ladder makes an angle $2 \tan^{-1} \mu$ with the vertical.

There are five forces on the ladder: its weight, and normal and frictional forces at A and B. But if you combine both pairs of normal and friction forces as single contact forces, the number of forces is reduced to three. Denote $\tan^{-1} \mu$ by λ, the angle of friction (see Section 7.3).

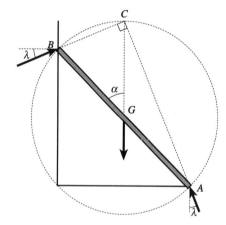

In Fig. 10.2, since friction is limiting, the contact force at A makes an angle λ with the vertical and the contact force at B makes an angle λ with the horizontal, so the two contact forces are at right angles to each other. Let their lines of action meet in C.

Fig. 10.2

The only other force is the weight of the ladder, so C must be directly above G, the centre of mass of the ladder. Also the ladder is uniform, so that G is the midpoint of AB.

The solution can be completed by a geometrical argument. Let the angle between the ladder and the vertical be α. A circle with centre G and diameter AB passes through C, so that triangle GCA is isosceles. Since angle $BGC = \alpha$ and angle $GCA = \lambda$, it follows that $\alpha = 2\lambda$, which is $2 \tan^{-1} \mu$.

EXAMPLE 10.1.3

A uniform roll of carpet of length l and weight W is lying on the floor. It is raised to the vertical by applying a force at one end which is always at right angles to the roll.

a Find the ratio of the frictional force to the normal contact force when the roll makes an angle θ with the floor.

b How large must the coefficient of friction be if the roll is not to slip on the floor while it is being raised?

a **Algebraic method 1** Fig. 10.3 shows the situation when the carpet roll makes an angle θ with the floor, being raised by a force P at right angles to its length. The normal force and the friction at the end on the floor are R and F.

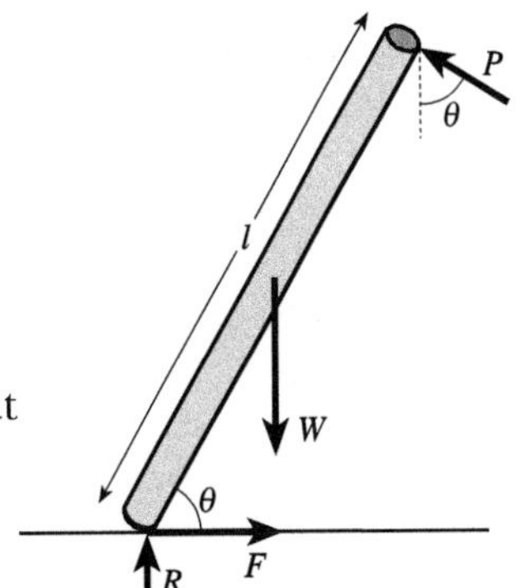

The simplest equation of moments to write is about the end of the roll in contact with the floor.

M (end on floor) $\quad Pl = W\left(\frac{1}{2}l\cos\theta\right)$,

which gives $\qquad P = \frac{1}{2}W\cos\theta$.

Since this equation doesn't involve either R or F, you will need two resolving equations to find them. Note that, since the roll makes an angle θ with the horizontal, the force P makes an angle θ with the vertical.

$\mathcal{R}(\rightarrow) \qquad\qquad F = P\sin\theta.$

$\mathcal{R}(\uparrow) \qquad R + P\cos\theta = W.$

From these you can find $F = \frac{1}{2}W\sin\theta\cos\theta$ and $R = W\left(1 - \frac{1}{2}\cos^2\theta\right)$, so

$$\frac{F}{R} = \frac{\sin\theta\cos\theta}{2 - \cos^2\theta}.$$

Fig. 10.3

Algebraic method 2 The drawback of Method 1 is that, by taking moments about the end of the roll on the floor, the equation doesn't involve either R or F, which are the two forces whose ratio you want to find. To get $\dfrac{F}{R}$ directly, you could take moments about the point X where the lines of action of the forces P and W intersect.

Then, with the notation of Fig. 10.4,

$M(X) \quad F \times XN = R \times XM.$

The disadvantage of this method is that, although XM is easily found as $XM = NO = \frac{1}{2}l\cos\theta$,

it is not so easy to find XN. For this you need the two right-angled triangles GQX and ONG:

$XN = XG + GN = \frac{1}{2}l\,\text{cosec}\,\theta + \frac{1}{2}l\sin\theta = \frac{1}{2}l\left(\text{cosec}\,\theta + \sin\theta\right).$

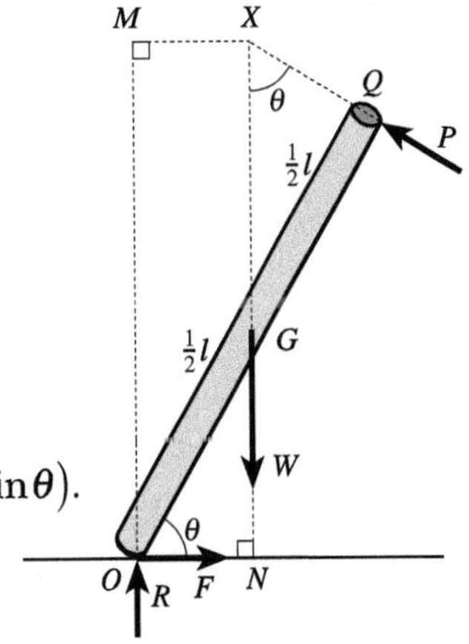

So

$$\frac{F}{R} = \frac{XM}{XN} = \frac{\frac{1}{2}l\cos\theta}{\frac{1}{2}l\left(\text{cosec}\,\theta + \sin\theta\right)} = \frac{\cos\theta}{\text{cosec}\,\theta + \sin\theta}.$$

Fig. 10.4

Geometrical method 1 Combine the friction and the normal contact force into a resultant contact force C at an angle α to the vertical, as in Fig. 10.5.

Then, since $\tan\alpha = \dfrac{F}{R}$, the required ratio can be found from the geometry of

Fig. 10.5. The solution uses the principle that the lines of action of P, W and C are concurrent.

With this method it is simpler to work with ϕ, the angle between the roll and the vertical, rather than with θ. There are two right-angled triangles in the figure, OQX and GQX, with a common side QX which can be found as either $OQ \tan \beta$ or as $GQ \tan \phi$. Since $OQ = l$ and $GQ = \frac{1}{2}l$,

$$l \tan \beta = \frac{1}{2} l \tan \phi, \qquad so \qquad \tan \beta = \frac{1}{2} \tan \phi$$

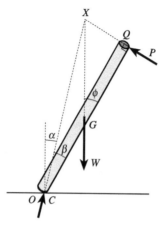

Fig. 10.5

Now you want to find $\tan \alpha$, and clearly $\alpha = \phi - \beta$. So

$$\tan \alpha = \tan(\phi - \beta) = \frac{\tan \phi - \tan \beta}{1 + \tan \phi \tan \beta} = \frac{\tan \phi - \frac{1}{2}\tan \phi}{1 + \frac{1}{2}\tan^2 \phi} = \frac{\frac{1}{2}\tan \phi}{1 + \frac{1}{2}\tan^2 \phi}.$$

To get the answer in terms of θ, note that $\phi = \frac{1}{2}\pi - \theta$, so $\phi = \dfrac{1}{\tan \theta}$. This gives

$$\tan \alpha \quad \frac{\dfrac{1}{2\tan \theta}}{1 + \dfrac{1}{2\tan^2 \theta}}.$$

Multiplying both numerator and denominator by $2 \tan^2 \theta$,

$$\frac{F}{R} = \tan \alpha = \frac{\tan \theta}{2 \tan^2 \theta + 1}.$$

Geometrical method 2 You could also use coordinates to find the angle α. Taking ON and OM in Fig. 10.4 as x- and y-axes, Q has coordinates $(l \cos \theta, l \sin \theta)$,

and QX has gradient $-\dfrac{1}{\tan \theta}$, so QX has equation $x \cos \theta + y \sin \theta = 1$. Also GX has equation $x = \frac{1}{2}l\cos\theta$, so the coordinates of X are $\left(\frac{1}{2}l\cos\theta, \dfrac{\frac{1}{2}l(2 - \cos^2 \theta)}{\sin \theta} \right)$. Then,

from Fig. 10.5, $\tan \alpha = \dfrac{x}{y} = \dfrac{\sin \theta \cos \theta}{2 - \cos^2 \theta}$.

b The various methods in part **a** give three apparently different expressions for $\dfrac{F}{R}$.

It is an interesting exercise in trigonometry to show that they are in fact the same.

If you draw its graph you will find that the ratio starts with the value 0 when $\theta = 0$, rises to a maximum and then falls back to 0 when $\theta = \frac{1}{2}\pi$. If the roll is not to slip as it is raised, the coefficient of friction must be greater than or equal to the maximum value.

The obvious way to find the maximum is to use calculus. For this you need to use the derivatives of various trigonometric functions and the rule for differentiating the quotient of two functions (see P2&3 Chapters 6 and 7). It is left to you to check that the maximum occurs when $\tan\theta = \sqrt{2}$, and that the maximum value is $\frac{1}{4}\sqrt{2}$.

You can, however, get the answer more quickly by using an algebraic method.

If the ratio $\dfrac{F}{R}$ is denoted by r, then the first geometric method in part **a** gives

$$r = \frac{\tan\theta}{2\tan^2\theta + 1},$$ which can be written as $2r\tan^2\theta - \tan\theta + r = 0$.

This is a quadratic equation whose solutions give the values of $\tan\theta$ for which r takes a particular value. For this equation to have roots, the discriminant

'$b^2 - 4ac$' must be positive or zero. That is $1 - 4(2r)r \geq 0$, which gives

$$r \leq \sqrt{\frac{1}{8}} = \frac{1}{4}\sqrt{2}.$$

So the ratio $\dfrac{F}{R}$ is never greater than $\frac{1}{4}\sqrt{2}$, which means that if $\mu \geq \frac{1}{4}\sqrt{2}$ the roll of carpet will not slip as it is raised to the vertical position.

Exercise 10A

Try to solve each question of this exercise by at least two methods.

1   A uniform rigid rod AB has length 2 m and weight 60 N. The rod is smoothly hinged at its end A to a vertical wall. The rod is held at an angle of 60° downward from the wall by a force of magnitude F N acting at B. The force acts at an angle of $\theta°$ upwards from the horizontal, as shown in the diagram.

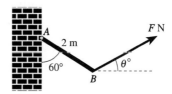

Show that $F\cos(60 - \theta)°$ is approximately 26 N.

Find the magnitude and direction of the force exerted by the wall on the rod at A when

   a $\theta = 30$,   **b** $\theta = 60$,
   c $\theta = 90$.

2   A uniform beam AB has length 4 m and mass 80 kg. A string is attached to the mid-point M of AB and passes over a small pulley P at the top of a wall. The beam is

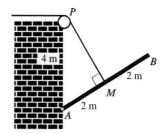

in equilibrium with A in contact with the wall at a point 4 m below P, as shown in the diagram. The string is taut with the part PM at right angles to AB. Find

a the tension in the string,

b the magnitude and direction of the force exerted by the wall on the beam at A.

3 A uniform lamina AOB is in the shape of a sector of a circle with centre O and radius 0.5 m, and has angle $AOB = \frac{1}{3}\pi$ radians and weight 3 N. The lamina is freely hinged at O to a fixed point and is held in equilibrium with AO vertical by a force of magnitude F N acting at B. The direction of this force is at right angles to OB (see diagram). Find

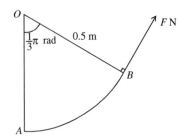

i the value of F,

ii the magnitude of the force acting on the lamina at O.

(Cambridge International AS and A Level Mathematics 9709/51 Paper 5 Q5 November 2009)

4 $ABCD$ is the central cross-section of a uniform rectangular box, with $AB = 90$ cm and $BC = 100$ cm. The face which includes DC rests against a step 14 cm above ground level, and the edge through D rests on the ground, 48 cm from the step. The contact with the step is smooth, and the contact with the ground is rough with coefficient of friction μ. If the box is on the point of slipping, find μ.

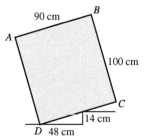

5 A uniform rod is in equilibrium at an angle θ to the vertical, with its lower end on the ground and its upper end resting against a small peg. The coefficient of friction μ is the same at both contacts. Prove that, if equilibrium is limiting,

$$\sin 2\theta = \frac{4\mu}{1 + \mu^2}.$$

10.2 Problems involving motion

The choice between algebraic and geometrical methods doesn't only occur in equilibrium problems. It also arises in problems which involve motion.

It was shown in M1 Section 10.1 that, if the sides of a vector force diagram are projected on a line in any direction, the relation between the projections is equivalent to an equation of resolving in that direction. The same reasoning can be applied to any vector relation.

For example, in Section 1.1 the displacement–time equation for a projectile was written in the vector form $\mathbf{r} = \mathbf{u}t + \frac{1}{2}\mathbf{g}t^2$. This is illustrated in Fig. 10.6. If you project the sides of this triangle on the x-axis and on the y-axis, you get the usual equations for the horizontal and vertical motion. But you can also project the sides in other directions, and this sometimes produces a neater solution to a problem.

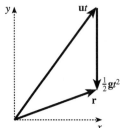

Fig. 10.6

EXAMPLE 10.2.1

Part of a golf course is on a hill which slopes at an angle α to the horizontal. The ball is hit straight up the hill with speed u at an angle θ to the horizontal.

a How far up the hill does it first land?

b For different values of θ, what is the greatest distance up the hill that the ball can be hit?

a Method 1 Fig. 10.7 shows a triangle OAB, which corresponds to the triangle in Fig. 10.6 applied to the flight of the golf ball. If t is the time that the ball is in the air, the sides of the triangle are $OA = ut$, $AB = \frac{1}{2}gt^2$ and $OB = r$.

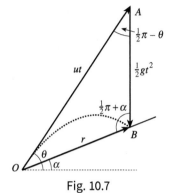

Fig. 10.7

The angles at O, A and B are $\theta - \alpha$, $\frac{1}{2}\pi - \theta$ and $\frac{1}{2}\pi + \alpha$. The sine rule then gives

$$\frac{r}{\sin\left(\frac{1}{2}\pi - \theta\right)} = \frac{ut}{\sin\left(\frac{1}{2}\pi + \alpha\right)} = \frac{\frac{1}{2}gt^2}{\sin(\theta - \alpha)},$$

or more simply

$$\frac{r}{\cos\theta} = \frac{ut}{\cos\alpha} = \frac{\frac{1}{2}gt^2}{\sin(\theta - \alpha)}.$$

From the first equality, $t = \dfrac{r\cos\alpha}{u\cos\theta}$; and from the second, $t = \dfrac{2u\sin(\theta - \alpha)}{g\cos\alpha}$. So

$$\frac{r\cos\alpha}{u\cos\theta} = \frac{2u\sin(\theta - \alpha)}{g\cos\alpha},$$

which gives

$$r = \frac{2u^2\sin(\theta-\alpha)\cos\theta}{g\cos^2\alpha}.$$

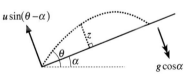

Fig. 10.8

Method 2 The simplest way to find the time that the ball is in the air is to resolve the velocity and acceleration vectors perpendicular to the slope, as shown in Fig. 10.8. The resolved part of the initial velocity is $u\cos\left(\frac{1}{2}\pi - (\theta - \alpha)\right)$, which can be written more simply as $u\sin(\theta - \alpha)$. The resolved part of the acceleration g is $-g\cos\alpha$. So, if z is the perpendicular distance of the ball from the slope at time t,

$$z = u\sin(\theta - \alpha)t - \frac{1}{2}(g\cos\alpha)t^2.$$

The ball lands when, $z = 0$, and this gives

$$t = \frac{2u\sin(\theta - \alpha)}{g\cos\alpha}.$$

Now resolve horizontally. The acceleration has no resolved part in this direction, and the velocity and displacement have resolved parts $u \cos \theta$ and $r \cos \alpha$. So $r \cos \alpha = (u \cos \theta)t$, which again gives

$$r = \frac{2u^2 \sin(\theta - \alpha) \cos \theta}{g \cos^2 \alpha}.$$

Method 3 Another approach is to use the standard cartesian equation for the trajectory (see Section 1.3(iv)). The point where the ball lands has coordinates $(r \cos \alpha, r \sin \alpha)$, and these must satisfy the equation. Therefore

$$r \sin \alpha = r \cos \alpha \tan \theta - \frac{g(r \cos \alpha)^2}{2u^2 \cos^2 \theta}.$$

Cancelling the common factor r and rearranging,

$$\frac{gr \cos^2 \alpha}{2u^2 \cos \theta} = \cos \theta (\tan \theta \cos \alpha - \sin \alpha).$$

The right side is $\sin \theta \cos \alpha - \cos \theta \sin \alpha = \sin(\theta - \alpha)$, so finally

$$r = \frac{2u^2 \sin(\theta - \alpha) \cos \theta}{g \cos^2 \alpha}.$$

b The formula for r can be written as the product of two parts,

$$r = \frac{u^2}{g \cos^2 \alpha} \times 2 \sin(\theta - \alpha) \cos \theta.$$

Only the second factor involves θ, and you can use the identity

$$2 \sin A \cos B \equiv \sin(A + B) + \sin(A - B)$$

(see P2&3 Example 5.3.4) to write this as

$$2 \sin(\theta - \alpha) \cos \theta = \sin(2\theta - \alpha) + \sin(-\alpha) = \sin(2\theta - \alpha) - \sin \alpha.$$

Now only the term $\sin(2\theta - \alpha)$ involves θ, and the greatest value of this is 1. So

$$r_{max} = \frac{u^2}{g \cos^2 \alpha} \times (1 - \sin \alpha).$$

This can be simplified by writing $\cos^2 \alpha$ as $1 - \sin^2 \alpha = (1 - \sin \alpha)(1 + \sin \alpha)$. You can then cancel the factor $1 - \sin \alpha$ to get

$$r_{max} = \frac{u^2}{g(1 + \sin \alpha)}.$$

Another equation which should properly be expressed as a vector relationship is Newton's second law of motion, $\mathbf{F} = m\mathbf{a}$. In this complete form, the equation states not just that the magnitude of the force is equal to the magnitude of the acceleration, but also that the force and the acceleration are in the same direction. The symbol $\mathbf{F}$ here stands for the resultant of all the forces acting on the object.

EXAMPLE 10.2.2

Cars race round a track which includes a horizontal circular arc of radius r. This stretch of the track is banked at an angle α to the horizontal, so that a car driven round the track at a constant speed v experiences no sideways frictional force.

a Find an equation connecting v, r, g and α.

b If the coefficient of friction is μ, what is the fastest speed at which a car can go round this stretch of the track without skidding?

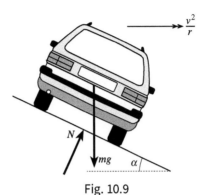

Fig. 10.9

   a **Method 1**   In this example the force which both supports the weight and provides the acceleration is the normal contact force N from the road (see Fig. 10.9). The acceleration is $\dfrac{v^2}{r}$, so for a car of mass m,

$$\mathcal{R}(\rightarrow) \qquad N\sin\alpha = m\frac{v^2}{r},$$

$$\mathcal{R}(\uparrow) \qquad N\cos\alpha - mg = 0.$$

Substituting $N = \dfrac{mg}{\cos\alpha}$ in the first equation, cancelling m and using $\dfrac{\sin\alpha}{\cos\alpha} = \tan\alpha$,

$$v^2 = rg\tan\alpha.$$

Method 2   Because the acceleration is in a horizontal direction, the resultant of the two forces mg and N must be horizontal. It follows that the force diagram for finding the resultant of mg and R has the form of Fig. 10.10. The resultant therefore has magnitude $mg\tan\alpha$, and Newton's second law takes the form

$$mg\tan\alpha = m\frac{v^2}{r}, \qquad \text{which give} \qquad v^2 = rg\tan\alpha.$$

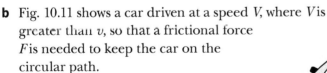

Fig. 10.10

b Fig. 10.11 shows a car driven at a speed V, where V is greater than v, so that a frictional force F is needed to keep the car on the circular path.

Here are three possible ways of proceeding.

Method 1   Resolve horizontally and vertically, as in part **a**.

$$\mathcal{R}(\rightarrow) \qquad N\sin\alpha + F\cos\alpha = m\frac{V^2}{r},$$

$$\mathcal{R}(\uparrow) \qquad N\cos\alpha - F\sin\alpha - mg = 0.$$

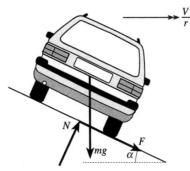

Fig. 10.11

These are simultaneous equations for N and F, which can be solved in the usual way. Multiplying the first equation by $\sin\alpha$, the second by $\cos\alpha$ and adding gives

$$(N\sin\alpha + F\cos\alpha)\sin\alpha + (N\cos\alpha - F\sin\alpha)\cos\alpha = \frac{mV^2}{r}\sin\alpha + mg\cos\alpha.$$

Using $\sin^2\alpha + \cos^2\alpha = 1$, the left side of this equation is just N, so

$$N = \frac{mV^2}{r}\sin\alpha + mg\cos\alpha.$$

Similarly, multiplying the first by $\cos\alpha$, the second by $\sin\alpha$ and subtracting leads to

$$F = \frac{mV^2}{r}\cos\alpha - mg\sin\alpha.$$

For the car not to skid, $F \leq \mu N$, so

$$\frac{mV^2}{r}\cos\alpha - mg\sin\alpha \leq \mu\left(\frac{mV^2}{r}\sin\alpha + mg\cos\alpha\right).$$

It is convenient to divide through the equation by $\cos\alpha$, and to use $\dfrac{\sin\alpha}{\cos\alpha} = \tan\alpha.$ Rearranging the inequality and dividing by m,

$$\frac{V^2}{r}(1 - \mu\tan\alpha) \leq g(\tan\alpha + \mu).$$

There are now two possibilities. If the road is so rough that $\mu\tan\alpha$ is greater than 1, the left side is negative and the right side is positive. In that case the inequality holds however large V is, so the car can round the bend at any speed.

Otherwise you can multiply the inequality by the positive quantity $\dfrac{r}{1 - \mu\tan\alpha}$ to get

$$V^2 \leq \frac{rg(\tan\alpha + \mu)}{1 - \mu\tan\alpha}.$$

The greatest speed at which a car can go round the curve is

$$\sqrt{\frac{rg(\tan\alpha + \mu)}{1 - \mu\tan\alpha}}.$$

Method 2   Instead of resolving parallel and perpendicular to the acceleration with the forces at an angle to these directions, you can resolve perpendicular to each unknown force in turn with the acceleration at an angle.

$$\mathcal{R}(\parallel \text{to the slope}) \; F + mg\sin\alpha = m\frac{V^2}{r}\cos\alpha,$$

$$\mathcal{R}(\perp \text{to the slope}) \qquad N - mg\cos\alpha = m\frac{V^2}{r}\sin\alpha.$$

These equations give

$$F = \frac{mV^2}{r}\cos\alpha - mg\sin\alpha, \qquad N = \frac{mV^2}{r}\sin\alpha + mg\cos\alpha,$$

as obtained by solving the simultaneous equations in Method 1. This method gets the expressions for F and N directly, and you can now complete the solution just as in Method 1.

Method 3 This method uses the angle of friction λ, where $\tan\lambda = \mu$.

Fig. 10.12 is derived from Fig. 10.11 by combining N and F into a total contact force C, which makes an angle θ with the normal to the slope. If the car is not to skid, then $\theta \le \lambda$.

The effect of this is to reduce the number of forces on the car to two, so that C is the force which both supports the weight and provides the acceleration. You can now see that Fig. 10.12 is essentially the same as Fig. 10.9, with two differences:

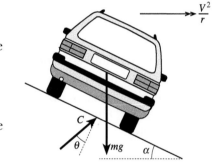

the speed v is replaced by V, and the force makes an angle $\alpha + \theta$, rather than α, with the vertical.

Fig. 10.12

So the result of part **a** can be adapted to give the equation

$$V^2 = rg\tan(\alpha + \theta).$$

Now $\tan(\alpha + \theta) \le \tan(\alpha + \lambda) = \dfrac{\tan\alpha + \tan\lambda}{1 - \tan\alpha\tan\lambda} = \dfrac{\tan\alpha + \mu}{1 - \mu\tan\alpha}.$

It follows that, for the car not to skid,

$$V^2 \le \frac{rg(\tan\alpha + \mu)}{1 - \mu\tan\alpha}.$$

Exercise 10B

Try to find the most efficient method to solve the problems in this exercise.

1 On a golf course, players have to hit the ball across a lake, starting with the ball on a tee. The far side of the lake is a distance a horizontally and $a\tan\alpha°$ vertically from the tee. If a player strikes the ball with initial speed u at an angle $\theta°$ to the horizontal, show that u has to satisfy the inequality

$$u^2 > \frac{ag\cos\alpha°}{\sin(2\theta + \alpha)° + \sin\alpha°}.$$

The far side of the lake is 150 metres horizontally from the tee and 20 metres below it.

a What value of θ will enable the player to clear the lake with the smallest possible initial speed?

b If the ball may leave the tee at any angle with the horizontal between $20°$ and $60°$, with what speed must the player hit it to be sure of clearing the lake?

2 A batsman B strikes a cricket ball, and it hits the pavilion clock C 14 metres above the level of the pitch at a horizontal distance of 48 metres. The initial direction of the trajectory makes an angle of $\tan^{-1}\frac{4}{3}$ with the horizontal. What is the angle between the initial direction and the line BC? Find

a how long the ball is in the air before it hits the clock,

b the speed with which the ball was struck,

c the greatest height above the pitch reached by the ball.

3 A hollow cone has base radius 4 metres and height 3 metres. It is mounted with its axis vertical and vertex pointing downwards, and a small pebble is placed on the inside surface of the cone at a distance of 2 metres from the vertex. In this position the pebble is in limiting equilibrium, about to slip down the cone towards the vertex. The cone is now set rotating about its axis. What is the greatest angular speed for which the pebble will remain in the same position on the surface of the cone?

4 A track for young racing drivers consists of a hollow in the shape of part of a sphere of radius 80 m. The total mass of a child and her car is 200 kg. At low speeds she drives round the track in horizontal circles banked at $\theta°$ to the horizontal; as the speed increases, so does θ. Find the speed when $\theta = 40$ if there is no sideways frictional force on the wheels.

As a safety procedure, a cord is attached to the side of the car and to an anchor at the lowest point of the hollow. When θ reaches the value 40 this cord becomes taut, so that the angle cannot get any bigger. Find the tension in the cord when the child is driving at 25 m s^{-1}.

5 The figure shows, in diagrammatic form, a governor for controlling the speed of an engine. This consists of a framework set in a vertical plane which rotates with the vertical shaft driven by the engine. A is a fixed point on the rotating shaft. Heavy spheres, each of mass m, are attached to the framework at B and D. C is a sleeve, of mass M, which can move up and down the shaft without friction. AB, BC, CD and DA are rods of equal length l and negligible mass, hinged at A, B, C and D so that the rhombus $ABCD$ can change its shape. When the shaft rotates with angular speed Ω, each rod is at an angle θ to the vertical. Find an expression for the tension in one of the lower rods in terms of m, l, Ω, θ and g, and deduce that $(M+m)g = ml\Omega^2 \cos\theta$.

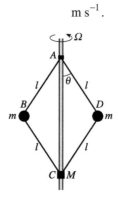

Miscellaneous exercise 10

1 A uniform beam of length 6 metres and mass 20 kg is hinged to a wall at one end O. A load of 10 kg is placed at the far end of the beam. The beam is supported at $60°$ to the upward vertical by the tension, T newtons, in a cable whose other end is fixed to the wall at A, 4 metres above the hinge. Where should the cable be attached to the beam for the magnitude of the force from the hinge to be as small as possible?

2 A uniform solid cone has height 30 cm and base radius r cm. The cone is placed with its axis vertical on a rough horizontal plane. The plane is slowly tilted and the cone remains in equilibrium until the angle of inclination of the plane reaches $35°$, when the cone topples. The diagram shows a cross-section of the cone.

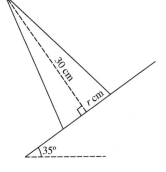

  i Find the value of r.

  ii Show that the coefficient of friction between the cone and the plane is greater than 0.7.

(Cambridge International AS and A Level Mathematics 9709/51 Paper 5 Q2 June 2010)

3 A projectile is launched from a point O with initial speed u at an angle θ to the horizontal. Its subsequent path is observed from O. At a certain instant it is seen at an angle of elevation α, and at time t later it is seen to land at the same level as O.

Show that $t = \dfrac{2u}{g}\cos\theta\tan\alpha$.

A cannonball was fired across a level plain. After 10 seconds it was seen at an angle of elevation of $20°$, and 5 seconds later it landed. Find the initial velocity of the cannonball.

4 ABC is a uniform triangular lamina of weight 19 N, with $AB = 0.22$ m and $AC = BC = 0.61$ m. The plane of the lamina is vertical. A rests on a rough horizontal surface, and AB is vertical. The equilibrium of the lamina is maintained by a light elastic string of natural length 0.7 m which passes over a small smooth peg P and is attached to B and C. The portion of the string attached to B is horizontal, and the portion of the string attached to C is vertical (see diagram).

  i Show that the tension in the string is 10 N.

  ii Calculate the modulus of elasticity of the string.

  iii Find the magnitude and direction of the force exerted by the surface on the lamina at A.

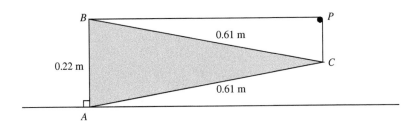

(Cambridge International AS and A Level Mathematics 9709/
51 Paper 5 Q5 June 2011)

5   A hemispherical bowl, of radius r, is fixed with its rim horizontal. A uniform rod of length l is in equilibrium at an angle α to the horizontal, resting against the rim of the bowl and with its lower end on the inside surface of the bowl. Prove that, if both contacts are smooth, then $l \cos \alpha = 4r \cos 2\alpha$.

6   A surface of revolution is formed by rotating a curve with equation $y = f(x)$, for $x > 0$, about the y-axis. The surface is modelled in a thin layer of smooth metal. A particle is placed on the surface at the point of the curve with coordinate x, and set in motion round the surface in a horizontal circle with constant angular speed Ω. Prove that $\Omega^2 = \dfrac{g f'(x)}{x}$.

   a   Obtain the equation of the curve $y = f(x)$ if the time T to complete one revolution is the same wherever the particle is placed.

   b   Obtain the equation of the curve if the speed V of the particle is the same wherever it is placed.

   c   The surface is used to demonstrate how the planets go round the sun. Kepler's third law states that, if x is the radius of an orbit (approximated as circular), then $\Omega^2 = \dfrac{c}{x^3}$, where c is constant. Obtain the equation of the curve needed to simulate this motion.

Revision exercise 2

1 A uniform triangular lamina ABC of weight 200 newtons is right angled at B and has $AB = 30$ cm. It is smoothly pivoted at A to a fixed point, and maintained in equilibrium with AB horizontal by a force of magnitude F newtons, which acts at 60° to the horizontal, as shown in the diagram. Find the value of F, and the magnitude and direction of the force exerted on the triangle by the pivot.

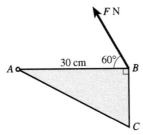

2 A golfer drives the ball with a speed of $40\,\text{m s}^{-1}$ at 20° to the horizontal. Find how far from the starting point it will land

  a if the ground is horizontal,

  b if the ground slopes down from the starting point at 5° to the horizontal.

What forces might act on the ball which would cast doubt on the accuracy of answers obtained by using the simple gravity model?

3 On a fairground there is a cylindrical chamber of radius 2 metres which can rotate about a vertical axis. Customers stand inside the chamber with their backs to the side wall. The coefficient of friction between their clothes and the wall of the chamber is 0.4. The chamber is then set rotating. When it is rotating fast enough, the floor of the chamber is suddenly removed, and the customers stay in position supported by friction. What angular speed is necessary for this to happen?

4 The acceleration that an electric car of mass $1200\,\text{kg}$ can produce is limited by two factors: the frictional force between the driving wheels and the road cannot exceed 6000 newtons, and the maximum power output of the engine is $30\,\text{kW}$. What is the shortest time in which it could reach a speed of $20\,\text{m s}^{-1}$ from a standing start?

5 A uniform ornamental stone has the form of half a circular cylinder, and weighs 150 newtons. The diagram shows the cross-section containing the centre of mass. The stone is in equilibrium with this cross-section in a vertical plane, with one end A of the diameter in contact with horizontal ground.

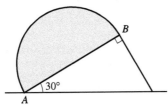

The other end B of the diameter is supported by a prop, which is in the same vertical plane as the cross-section and at right angles to AB. The diameter AB makes an angle of 30° with the horizontal. Find

  a the magnitude of the force exerted by the prop on the stone,

  b the magnitude and direction of the force exerted by the ground on the stone.

6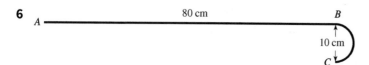

A thin walking stick has a straight section AB 80 cm long, and a semicircular handle BC with diameter 10 cm. Locate the point X of AB about which the stick will balance with the straight section horizontal.

The stick is now hung over a shelf with the end C resting on the shelf. What angle will AB then make with the vertical?

7 A particle P is projected with speed $V\,\mathrm{m\,s^{-1}}$ at an angle of 30° above the horizontal from a point O on horizontal ground. At the instant 2 s after projection, OP makes an angle of 15° above the horizontal. Calculate V.

(Cambridge International AS and A Level Mathematics 9709/51 Paper 5 Q1 November 2014)

8 A uniform solid body has a cross-section as shown in Fig. I.

i Show that the centre of mass of the body is 2.5 cm from the plane face containing OB and 3.5 cm from the plane face containing OA.

ii The solid is placed on a rough plane which is initially horizontal. The coefficient of friction between the solid and the plane is μ.

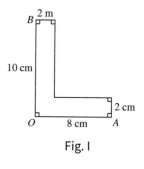

Fig. I

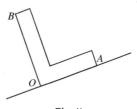

Fig. II

a The solid is placed with OA in contact with the plane, and then the plane is tilted so that OA lies along a line of greatest slope with A higher than O (see Fig. II). When the angle of inclination is sufficiently great the solid starts to topple (without sliding). Show that $\mu > \frac{5}{7}$.

b Instead, the solid is placed with OB in contact with the plane, and then the plane is tilted so that OB lies along a line of greatest slope with B higher than O (see Fig. III). When the angle of inclination is sufficiently great the solid starts to slide (without toppling). Find another inequality for μ.

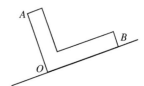

Fig. III

(Cambridge International AS and A Level Mathematics 9709/05 Paper 5 Q7 June 2009)

9 A particle P of mass m kg is attached to one end of a light inextensible string of length L m. The other end of the string is attached to a fixed point O. The particle P moves with constant speed in a horizontal circle, with the string taut and inclined at 35° to the vertical. OP rotates with angular speed $2.2\,\mathrm{rad\,s^{-1}}$ about the vertical axis through O (see diagram). Find

i the value of L,

ii the speed of P in $\mathrm{m\,s^{-1}}$.

(Cambridge International AS and A Level Mathematics 9709/05 Paper 5 Q3 June 2006)

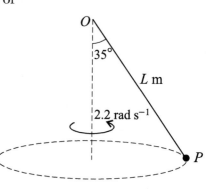

10 A small bead B of mass m kg moves with constant speed in a horizontal circle on a fixed smooth wire. The wire is in the form of a circle with centre O and radius 0.4 m. One end of a light elastic string of natural length 0.4 m and modulus of elasticity $42m$ N is attached to B. The other end of the string is attached to a fixed point A which is 0.3 m vertically above O (see diagram).

  i Show that the vertical component of the contact force exerted by the wire on the bead is $3.7m$ N upwards.

  ii Given that the contact force has zero horizontal component, find the angular speed of B.

  iii Given instead that the horizontal component of the contact force has magnitude $2m$ N, find the two possible speeds of B.

The string is now removed. B again moves on the wire in a horizontal circle with constant speed. It is given that the vertical and horizontal components of the contact force exerted by the wire on the bead have equal magnitudes.

  iv Find the speed of B.

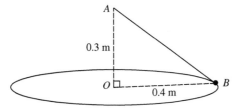

(Cambridge International AS and A Level Mathematics 9709/51 Paper 5 Q7 June 2014)

11 An ancient civilisation constructed monuments in the shape of circular cones. Some of these have fallen over, and are lying on their sides on level ground. An archaeologist wants to move these, and the first step is to place small rollers underneath them. To do this, a rope is attached to the highest point of the rim, and this is pulled horizontally by a team of student volunteers so as to lift either

  a the vertex, or **b** the rim of the monument off the ground. If in the fallen position the base of the monument is at an angle $\alpha°$ to the horizontal, find in terms of α how to decide which of **a** or **b** will require less effort.

12 A particle P is projected with speed 30 m s^{-1} at an angle of 60° above the horizontal from a point O on horizontal ground. For the instant when the speed of P is 17 m s^{-1} and increasing,

  i show that the vertical component of the velocity of P is 8 m s^{-1} downwards,

  ii calculate the distance of P from O.

(Cambridge International AS and A Level Mathematics 9709/51 Paper 5 Q5 November 2012)

13 A bird-lover suspends a half-coconut of mass $\frac{1}{2}$ kg from a tree by a string attached to a point of the rim. Modelling the half-coconut as a hemispherical shell of radius r, find the angle which the plane of the rim makes with the vertical.

A bird now perches on the lowest point of the rim, and as a result the angle is reduced to 20°. Find the mass of the bird.

14 A hollow cone is fixed with its vertex O downwards and its axis vertical. A particle of mass m on the smooth inner surface of the cone moves round it in a horizontal circle whose centre is at a height h above O. Prove that it has kinetic energy $\frac{1}{2}mgh$.

15 The figure shows a soap dispenser for use in a washroom. It consists of a hemisphere of radius 3 cm joined at its rim to a cone of height 4 cm. It is made of thin metal of uniform thickness, and it can pivot about a horizontal axis which passes through its centre of mass. Find how far this axis is below the level of the rim of the hemisphere.

The dispenser is now filled with liquid soap up to the level of the nozzle at the vertex of the cone. (Users tilt the dispenser upside-down so that soap comes out of the nozzle.)

Show that the centre of mass of the dispenser and its contents is below the axis.

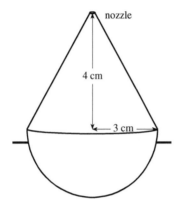

16 A particle P of mass 0.1 kg is projected vertically upwards from a point O with speed $20\,\text{m}\,\text{s}^{-1}$. Air resistance of magnitude $0.1v\,\text{N}$ opposes the motion, where $v\,\text{m}\,\text{s}^{-1}$ is the speed of P at time $t\,$s after projection.

i Show that, while P is moving upwards, $\dfrac{1}{v+10}\dfrac{\text{d}v}{\text{d}t} = -1$.

ii Hence find an expression for v in terms of t, and explain why it is valid only for $0 \le t \le \ln 3$.

iii Find the initial acceleration of P.

(Cambridge International AS and A Level Mathematics 9709/51 Paper 5
Q7 November 2009)

17 A uniform cuboidal column has a square base $ABCD$ with side 1 metre and height 3 metres. It is lying with the oblong face containing AB on the ground. The coefficient of friction between the column and the ground is μ. An attempt is made to raise the column in a series of short stages into a vertical position. A rope is attached to the mid-point of CD, and at each stage a gradually increasing horizontal force is applied in a direction perpendicular to CD. Show that the attempt cannot begin to succeed unless μ is greater than 1.5; but that if the first stage is successful, then it will be possible to raise the column until its centre of mass is directly above AB.

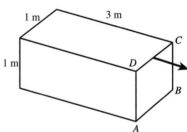

The rope is then replaced by a pole, and a force is applied in the opposite direction to slow the rotation of the column into the vertical position. Show that this will be successful to start with, but that eventually the column will begin to slip on the ground. Find the angle θ that the column will make with the vertical when this occurs, in terms of μ.

18 A particle P is projected from a point O on horizontal ground with speed $V\,\mathrm{m\,s^{-1}}$ and direction $60°$ upwards from the horizontal. At time t s later the horizontal and vertical displacements of P from O are x m and y m respectively.

   i Write down expressions for x and y in terms of V and t and hence show that the equation of the trajectory of P is

$$y=\left(\sqrt{3}\right)x-\frac{20x^2}{V^2}.$$

   P passes through the point A at which $x = 70$ and $y = 10$. Find

   ii the value of V,

   iii the direction of motion of P at the instant it passes through A.

          (Cambridge International AS and A Level Mathematics 9709/05 Paper 5
          Q7 November 2008)

19 A small aircraft, of weight 8000 newtons, is flying over an airfield at speed $U\,\mathrm{m\,s^{-1}}$. The aircraft propeller produces a thrust force T newtons in the direction of motion. The wing produces a lift force L newtons, which always acts at right angles to the thrust. The thrust T needed to maintain flight speed U varies with

U and L according to the equation $T = 0.1U^2 + \dfrac{0.004L^2}{U^2}$. Show that the aircraft

speed in straight and level flight which requires minimum T is $40\,\mathrm{m\,s^{-1}}$.

The pilot increases the propeller thrust by 40% from this minimum value. He banks the aircraft so that it starts to move in a horizontal circle of radius R metres with speed $32\,\mathrm{m\,s^{-1}}$. Use the thrust equation to find the lift force now provided by the wing. Hence deduce the inwards acceleration of the aircraft as it flies round the circle and determine R.

20 $OABC$ is the cross-section through the centre of mass of a uniform prism of weight 20 N. The cross-section is in the shape of a sector of a circle with centre O, radius $OA = r$ m and angle $AOC = \frac{2}{3}\pi$ radians. The prism lies on a plane inclined at an angle θ radians to the horizontal, where $\theta < \frac{1}{3}\pi$. OC lies along a line of greatest slope with O higher than C.

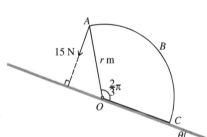

The prism is freely hinged to the plane at O. A force of magnitude 15 N acts at A, in a direction towards the plane and at right angles to it (see diagram). Given that the prism remains in equilibrium, find the set of possible values of θ.

          (Cambridge International AS and A Level Mathematics 9709/51 Paper 5
          Q7 June 2013)

Practice exam-style paper 1

Time 1 hour 15 minutes

Answer all the questions.

The use of an electronic calculator is expected, where appropriate.

1 The diagram shows a uniform triangular lamina ABC with $AB = 7$ cm, $BC = 24$ cm and angle $ABC = 90°$. The lamina is pivoted at A so that it can rotate freely in a vertical plane. The lamina is held in equilibrium, with BC horizontal and B vertically below A, by means of a force of magnitude FN acting at C perpendicular to AC and in the plane of the lamina. The weight of the lamina is WN. Find F in terms of W. [4]

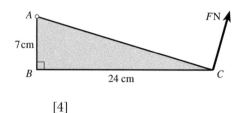

2 One end of a light elastic string is attached to a fixed point O, and a particle hangs in equilibrium at the other end A. If the mass of the particle is 0.1 kg the length of OA is 1.2 m. If the mass is 0.3 kg the length is 1.3 m. Find the modulus of elasticity of the string. [5]

3 A particle P of mass 0.5 kg is attached by light elastic strings to points A and B which are at the same horizontal level and a distance 1.2 m apart. Each string has natural length 0.6 m and modulus of elasticity 30 N. The particle is projected vertically downwards with speed v m s^{-1} from the mid-point of AB (see diagram). Given that P comes to instantaneous rest at a distance of 0.8 m below the level of AB, find v. [6]

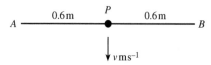

4 A uniform solid has cross-section $ABCDEF$ with shape and dimensions as shown in the diagram; all the angles are right angles. The centre of mass of the solid is a distance x cm from AF.

i Show that $x = \dfrac{20 + 4d + d^2}{20 + 2d}$. [4]

ii The solid is placed with AB on horizontal ground, and remains in equilibrium. Find the greatest possible value of d. [3]

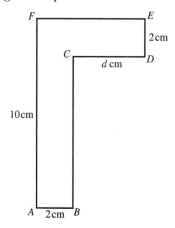

5 When a gun is fired, the shell is propelled along the barrel by the pressure of the gas generated by the explosive charge. When the shell is at a distance x m from the closed end of the barrel, the speed is v m s^{-1} (see diagram). In a simple model of the motion, the force exerted by the gas on the shell is taken to be proportional to $x^{-\frac{3}{2}}$, and all resistances to the motion of the shell are neglected. The barrel is assumed to be horizontal.

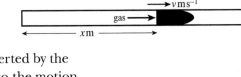

i Write down a differential equation relating v and x, and hence show that

$$v^2 = A - \frac{B}{\sqrt{x}},$$

where A and B are constants. [3]

ii The shell starts from rest at a distance of 0.25 m from the closed end of the barrel. The shell emerges from the barrel, which is 2.25 m long, at a speed of 400 m s^{-1}. Find the initial acceleration of the shell. [5]

6 The figure shows a carousel at a fairground. Children are strapped into chairs at the end of poles 4 metres long. The poles are hinged at the top to a horizontal drum of radius 3 metres, which rotates about a vertical axis. When the carousel is rotating at full speed, the poles swing out at 20° to the vertical, in vertical planes through the axis of rotation. The mass of the poles is small, and can be neglected. Calculate the time that the carousel takes to make one complete revolution at full speed. [8]

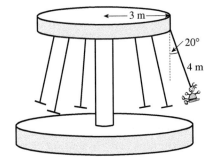

7 An anti-aircraft gun fires shells with initial velocity 400 m s^{-1} at an angle $\theta°$ above the horizontal, and the shells may be assumed to move freely under gravity. The target is a pilotless aircraft which flies at a speed of 100 m s^{-1} directly towards the gun at a constant height of 3500 m. A shell fired from the gun hits the aircraft when it is at a horizontal distance of x m from the gun.

i By using the equation of the trajectory of the shell, or otherwise, show that

$$x^2 \tan^2 \theta° - (3.2 \times 10^4) x \tan \theta° + (x^2 + (1.12 \times 10^8)) = 0$$ [4]

ii Hence find the greatest value of x at which the aircraft can be hit by a shell. [2]

iii For a hit at the greatest possible value of x, find

a the angle to the horizontal at which the shell should be fired, [2]

b the angle of elevation of the aircraft from the gun at the instant the shell is fired. [4]

199

Practice exam-style paper 2

Time 1 hour 15 minutes

Answer all the questions.

The use of an electronic calculator is expected, where appropriate.

1 A uniform semicircular lamina has radius 10 cm.

  i State, in terms of π, the distance of the centre of mass of the lamina from the midpoint of its diameter. [1]

  ii The lamina is freely suspended from one end of its diameter, and hangs in equilibrium. Find the angle that its diameter makes with the vertical. [2]

2 The point O is a distance of 0.3 m above a smooth horizontal plane. A particle P of mass 0.1 kg lies on the plane and is connected to O by means of a light inextensible string of length 0.5 m. The particle moves in a horizontal circle, in contact with the plane, with angular speed ω rad s^{-1} (see diagram).

  i Show that the tension in the string is $0.05\omega^2$ N. [3]

  ii Find the greatest value of ω for which contact between P and the plane is maintained. [3]

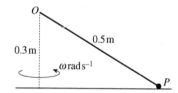

3 The diagram shows a uniform lamina $ABCD$ in which $AB = 2$ m, $AC = CD = 1$ m and angles BAC and ACD are right angles. The centre of mass of the lamina is at the point G. Show that the distance of G from AC is $\frac{1}{3}$ m, and find its distance from AB. [6]

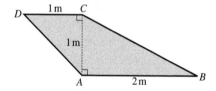

4 A and B are fixed points, with A a distance of 3 m vertically above B. A particle P of mass 1 kg is attached to A and B by means of two light elastic strings, each of natural length 1 m and modulus of elasticity 20 N (see diagram).

  i P hangs in equilibrium between A and B, with both strings taut. Find the distance AP. [3]

  ii P is released from rest at B. Find the speed of P when it is at a height of 2 m above B. [4]

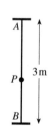

5 A fire-fighter is standing 25 m away from a building that is on fire. He directs a jet of water so that it passes horizontally through a window of the building which is 11 m above the ground. The jet is delivered from a hose at a height of 1 m above the ground. As it leaves the hose, the water jet has velocity components $U \text{m s}^{-1}$ horizontally towards the building and $V \text{m s}^{-1}$ vertically upwards.

   i Find V, and show that a drop of water takes approximately 1.4 s to travel from the hose to the window. [4]

   ii Find U, and hence find the speed and direction of the water as it leaves the hose. [4]

6 A car of mass 750 kg travels along a straight horizontal road. Its engine has constant power 20 kW and the resistance to motion has magnitude kv N, where k is a constant and $v \text{m s}^{-1}$ is the car's speed.

   i The greatest steady speed that the car can maintain is 80m s^{-1}. Find the value of k. [2]

   ii Show that the car's equation of motion can be expressed in the form

$$\frac{2v}{6400-v^2}\frac{\mathrm{d}v}{\mathrm{d}t}=\frac{1}{120}.$$ [3]

   iii Given that the car starts from rest at time $t = 0$, find v in terms of t.

7 A uniform beam AB has length $4l$ and weight W. It rests in equilibrium in contact with a small fixed peg P, where $AP = 3l$, and with the end A on horizontal ground. The beam makes an angle α with the ground (see diagram).

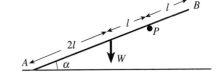

   i Suppose that the contact at A is smooth and the contact at P is rough.

     a Find the normal reaction acting on the beam at P. [2]

     b The coefficient of friction at P is μ_P. Show that $\mu_p \geq \tan\alpha$. [2]

   ii Suppose instead that the contact at A is rough, with coefficient of friction μ_A, and the contact at P is smooth. Show that

$$\mu_A \geq \frac{2\sin\alpha\cos\alpha}{3-2\cos^2\alpha}.$$ [7]

Answers

Most non-exact numerical answers are given correct to 3 significant figures.

1 The motion of projectiles

Exercise 1A (page 4)

1   47 m, 2.45 s

2   3.5 m

3   1.3 s

4   39 m at 33° below the horizontal 25 m s⁻¹ at 62° below the horizontal

5   0.58 s, 6.9 m s⁻¹

6   475 m

7   46.7 m s⁻¹

8   48.6°

9   21.2 m s⁻¹ at 70.7°; 28 m

10   **b** 1.3 s, 2.7 s
    c 1.27 s, 2.72 s

Exercise 1B (page 10)

1   13 m s⁻¹, 20 m s⁻¹ downwards

2   72 m, 28.8 m

4   $\begin{pmatrix} 30 \\ -20 \end{pmatrix}$ m

5   80.4 m, 22.5 m; 26.8 m s⁻¹, 7.5 m s⁻¹ down

6   **a** 7 s
    b 70 m s⁻¹
    c 490 m

7   **a** 0.4 s   **b** 4 m s⁻¹
    c 14.9 m s⁻¹   **d** 5.97 m

8   **a** 20 m s⁻¹
    b 15 m s⁻¹ 25 m s⁻¹, 36.9° above the horizontal

9   36.0°

10   28.3 m s⁻¹, 24.5 m s⁻¹

11   19.5 m s⁻¹

12   1.65 s

13   14.5 m

14   **a** 16.3 m s⁻¹, 42.5°
    b 2 m
    c 6.05 m

15   **i**   Speed: 13.5 m s⁻¹, $T = 0.721$
    ii 2.85°

16   **a** $\begin{pmatrix} 15 \\ -3 \end{pmatrix}$ m s⁻¹
    b 4 m, 1.2 m

17   24.6 m s⁻¹ at 54.3° above the horizontal

18   **i**   1.2 s
    iii 13.6 m

19   **i**   9.14 m s⁻¹, 40(.0)° below horizontal
    ii 2.58 s

20   3.55 s, 75.5 m

Exercise 1C (page 16)

1   17.8°, 72.2°

2   113 m s⁻¹

3   2.57 s, 39.4 m

4   11.6 m s⁻¹

5   69.0°

6   8.14 m s⁻²

7   375 m, 20.2 m, 325 m

8   36.9 m s⁻¹

9   32.0° (or 68.2°)

10   17.7 m s⁻¹

11   35.9°

12   31.1 m s⁻¹

13   21.8°

Miscellaneous exercise 1 (page 17)

1   1.93 s

2   **i** $x = (20 \cos 45)t$, $y = (20 \sin 45)t - \dfrac{gt^2}{2}$
    iii 40 m

3   **ii** 0.3(00) s

4   **ii** $V \sin \theta = 10$, $\theta = 22.6°$, $V = 26$

5   **i**   36.9°, 19.8 m s⁻¹
    ii 37.5 m
    iii 7.03 m

6   **i**   10 m s⁻¹
    ii $y = 15$ (below) or −15, $v = 20$ m s⁻¹

7   **b** $\dfrac{4\sqrt{3}k^2}{9g}$

8   **i**   3.8 m
    ii 1.89 s, 29(.0) m

9 i $x = (10 \cos 45°)t$ and $y = (10 \sin 45°)t - \dfrac{gt^2}{2}$;

$y = x - \dfrac{x^2}{10}$

ii 4.23

iii 8.79°

10 i $x = (15 \cos 41)t$, $y = (15 \sin 41)t - \dfrac{gt^2}{2}$

ii $H = 2.82\,\text{m}$, $D = 22.5\,\text{m}$

2 Moments

Exercise 2A (page 26)

1 27.8 N

2 5.04 m

3 28 kg, 32 kg; 60 N

4 46 cm

5 a 48

b 4

c 360

6 a no

b yes

7 1.5 m

Exercise 2B (page 30)

1 a 258 N   **b** 682 N

2 8240 N, 12 560 N

3 13 N, 1167 N; 2 kg

4 9 m

5 56.6 N, 73.4 N; assume the weight acts in a line midway across the curtain's 'spread', 0.8 m

6 125 N, 125 N; 112.2 N, 117.8 N

7 a 630 N

b 283.5 N

c 3.15 m s⁻²

Exercise 2C (page 35)

1 a 360 N m anticlockwise

b 384 N m anticlockwise

c 385 N m anticlockwise

d 170 N m anticlockwise

2 a 210 N m clockwise

b 280 N m anticlockwise

c 200 N m anticlockwise

d 900 N m clockwise

3 a 326 N m anticlockwise

b 247 N m anticlockwise

c 333 N m anticlockwise

d 126 N m anticlockwise

4 a 10   **b** 12

c 8

5 a 30   **b** 165.5

c 9.6   **d** 172.8

6 10 800 N

7 80 N

8 23.8 N

9 7.28 N

10 49.9

11 708 N

13 a 60 N

b 129 N

c 54.4 N

14 34.7 N

a 25 N

b 4.41 N

c 4.34 N

Miscellaneous exercise 2 (page 38)

1 38 cm

2 $R_A = 500\,\text{N}$, $R_B = 300\,\text{N}$

3 $W = 180$, $x = 3.3$

4 75 N

5 a 306

b less: the distance of the line of action of the force from A will be greater, so the force needed to produce the same moment is less.

6 i 18.6   **ii** 21.8

3 Centre of mass

Exercise 3A (page 45)

1 40 cm

2 0.47 m

3 4610 km

4 13.0 cm

5 33.1 cm

6 106.5 cm

7 16.0 cm

8 0.83 m

9 5 cm from AB; 2.85 cm from AC

10 **a** 5 cm

    b 7.5 cm

11 19.2 cm

12 0.934

13 **a** 8 cm

    b 6.4 cm

14 39.5 cm, 28.6 cm

15 2.1, 5.55

16 (6.86 cm, 3.74 cm)

Exercise 3B (page 54)

2 **a** Vertical, with A above B

    b Vertical, with B above A

    c At any angle

3 **a** 0, 10 N

    b 6.67 N, 3.33 N

5 $53.1 < \alpha < 126.9$

6 **a** 56.3°   **b** 25.5°

    c 38.0°   **d** 57.4°

7 **a** 51.0°

    b 5.7°

8 21.2°

9 **a** 23.0°

    b 2.0°

10 **a** P   **b** S

    c R   **d** Q

11 Topples; not the same (that is, it does not topple)

Miscellaneous exercise 3 (page 57)

1 **a** 5.58 cm   **b** 2.08 cm

2 **a** 2 cm   **b** 3 cm

3 56.4 cm

4 From vertical: 0.129 m; from horizontal 0.872 m

5 $\frac{1}{3} H$

6 24.4°

7 **a** 37.625

    b 13.625

8 55.0°

9 $h = 3.64$ cm, $\mu = 0.364$

10 $\tan\theta = \dfrac{M}{M + 2m}$

11 **a** **i** 3.92 cm

       ii 3.08 cm

    b max α is 38.2°

13 **i** 7.5 m   **ii** 0.417 W

14 **i** 0.22 m   **ii** 30

4 Rigid objects in equilibrium

Exercise 4A (page 67)

1 $600x$ N

2 0.588

3 **a** 100 N, 400 N, 100 N

    b 97.5 N, 475 N, 62.7 N

4 237; 0.297

5 **a** 672 N   **b** 955 N

6 0.275; 31.4 N, 104 N; 54.1 N, 73.2°

7 71.0 N, 37.6 N, 0.530

8 $\frac{22}{25}, \frac{7}{25} W$; 2.2

Exercise 4B (page 75)

1 **a** 20°   **b** 21.8°; by toppling

2 Between 0.2 and 0.3

3 More than 0.276 m from the back of the sledge

4 **a** 25 N   **b** Greater than 0.174

5 $\tan\alpha > 2 - \dfrac{1}{\mu}$; yes

7 100, 6

8 **a** 13

    b 67.4°

    c $8\frac{2}{3}$ m

9 75 N, 25 N; $3y = x + L$

10 **a** $\begin{pmatrix} -2 \\ -4 \end{pmatrix}$, $y = 2x$

    b $\begin{pmatrix} -4 \\ -1 \end{pmatrix}$, $y = \frac{1}{4}x + 4$

    c $\begin{pmatrix} 0 \\ 2 \end{pmatrix}$, $x = 3$

    d The given forces are already in equilibrium.

    e It is impossible to achieve equilibrium with a single force.

11 24.5 cm

12 **a** $G + R = W$, $S = F$, $W + 4F = 2R$, $4S + 2G = W$

    b $4S + W = 2R$, no

    c **i** no   **ii** yes

    d $\sqrt{5} - 2$

Miscellaneous exercise 4 (page 78)

1 $y = 2x - 20$

2 **i** 12 N

    ii 0.289

3 **i** 2.12 N

    ii 0.313

4 **i** $\theta = 47(.0)°$

    ii $\mu = 0.532$

5 **i** 960 N

    ii $X = 268.8$, $Y = 521.6$

6 **i** 562.5 N

    ii $\alpha = 52.1°$ (with vertical) or $\theta = 65.99°$ (with beam)

9 The trough falls first.

10 **a** $\frac{1}{5}\sqrt{3}W, \frac{1}{2}\sqrt{3}W, \frac{7}{10}W, \frac{9}{20}W$

11 **ii** 0.424

12 **b** W **c** 0.248

13 **i** 7.5

5 Elastic strings and springs

Exercise 5A (page 91)

1 3.125×10^{8} N

2 28.8 cm

3 **a** 0.192 m   **b** 0.461 m

4 24 cm

5 100 N

6 $1\frac{1}{3}$ N

8 $\dfrac{\sqrt{3}\lambda}{l}(2x - l)$

9 **a** 21 cm

    b 15 cm

10 **a** $\dfrac{2\lambda x}{l}$

    b $\dfrac{\lambda(l + x)}{l}$

    c $\dfrac{2\lambda(x - l)}{l}$

11 2λ

12 $x(y - 1) = $ constant

13 0.976 m

Exercise 5B (page 97)

1 735 J, 794 J

2 22.4 m s^{-1}, 27 950 J

3 7.5 m s^{-1}; 2.5 m (on the other side of O); the particle oscillates through 2.5 m on either side of O.

4 7.5 m s^{-1}; 1.5 m; the particle oscillates between 1.5 m and 2.5 m on the same side of O.

5 5.66 m s^{-1}

6 **a** 3.3 J   **b** 2.90 m s^{-1}

7 $\frac{1}{2}mgx$

8 9.6 kN

9 1.73 m

10 20.5 m s^{-1}

11 12.6 m s^{-1}

12 300 N, 8.91 m s^{-1}

13 **a** $\sqrt{u^{2} + 2gx - \dfrac{\lambda x^{2}}{ml}}$

    b $\dfrac{mgl + \sqrt{ml\left(ml^{2}l + \lambda u^{2}\right)}}{\lambda}$

    c The resultant force is downwards/upwards when the particle is above/below the equilibrium position.

14 $mg\left(l + \frac{1}{2}a\right)$

Miscellaneous exercise 5 (page 98)

1 **i** 4 N

    ii 8 m s^{-2}

2 4 m s^{-2}

3 $x = 0.9$

4 **i** 70 m s^{-2}

    ii 12 m s^{-1}

5 **ii** 2.83 m s^{-1}

    ii 2.45 m s^{-1}

6 1.11 m s^{-1}

7 **i** 0.2 kg

    ii (−)3.27 m s^{-2}

8 **ii** 2.75 m s^{-1}

9 ii $5.30\,\mathrm{m\,s^{-1}}$

   iii $30\,\mathrm{m\,s^{-2}}$ (upwards)

10 i $12\,\mathrm{N}, 24\,\mathrm{N}$

   iii $7.5\,\mathrm{m\,s^{-2}}$

   iv 0.5

11 i $6.25\,\mathrm{N}$

   ii $4.90\,\mathrm{m\,s^{-1}}$

Revision exercise 1 (page 102)

1 $1.3\,\mathrm{m}$

2 a $6000\,\mathrm{N}$

  b $30\,000\,\mathrm{N}, 45\,000\,\mathrm{N}$

3 $35.5\,\mathrm{m}$

4 $l > 45$

5 ii $8\,\mathrm{m}$

   iii $6.71\,\mathrm{m\,s^{-1}}$

6 a $-\frac{1}{4}mgl$

  b $-\frac{1}{2}mgl$; $\sqrt{\frac{1}{2}gl}$

7 a 2.92

  b 1.13

8 i $1.22\,\mathrm{m}$

   ii $6\,\mathrm{m\,s^{-1}}$

   iii $60\,\mathrm{m\,s^{-2}}$

9 ii For $\tan\theta = 0.75$, distance is $38.4\,\mathrm{m}$. For $\tan\theta = 4.25$, distance is $17.8\,\mathrm{m}$.

   iii A sketch of two parabolic arcs which intersect once. Both start at the origin, each with $y \geq 0$ throughout, and each returns to the x-axis. The arc with the smaller angle of projection should have the greater range. The range of the arc with smaller angle of projection should appear significantly greater than the value x at the intersection. The range of the other arc should be slightly greater than x at the intersection.

10 a $7\,\mathrm{cm}$

  b $8.2\,\mathrm{cm}$

11 The left finger slips when the force is $0.375\,\mathrm{N}$; the force gradually increases to $0.5\,\mathrm{N}$, when the left finger reaches the $20\,\mathrm{cm}$ mark. Then both fingers move symmetrically, with no further increase in force, until they meet at the $50\,\mathrm{cm}$ mark.

   Up to the $69.2\,\mathrm{cm}$ mark

12 $800; 373\,\mathrm{N}$

13 i $33.3\,\mathrm{m\,s^{-1}}, 25.8°$

   ii $\theta = 19.3°$ with the horizontal

14 a It can be up to $1.6\,\mathrm{m}$ above the ground, at up to $0.3\,\mathrm{m}$ from the axis.

  b i $36.9°$ ii $23.6°$

15 ii $0.38(0)$

16 The tension increases from $\frac{1}{2}\sqrt{2}W$ when $\theta = 0$ up to about $0.79\,W$ when $\theta \approx 28$, then decreases to $\frac{1}{2}W$ when $\theta = 90$.

6 Motion round a circle

Exercise 6A (page 110)

1 $1.5\,\mathrm{rad\,s^{-1}}$

2 $0.449\,\mathrm{rad\,s^{-1}}$

3 $1.70\,\mathrm{rad\,s^{-1}}, 0.849\,\mathrm{m\,s^{-1}}$

4 $463\,\mathrm{m\,s^{-1}}$

5 $2\,\mathrm{m}$

6 $0.262\,\mathrm{rad\,s^{-1}}, 26.7\,\mathrm{m}$

7 $0.654\,\mathrm{rad\,s^{-1}}, 0.916\,\mathrm{rad\,s^{-1}}$

8 $8.17\,\mathrm{rad\,s^{-1}}, 1.23\,\mathrm{m\,s^{-1}}$

9 $12\,\mathrm{rad\,s^{-1}}$

10 $2.12\,\mathrm{cm}$

Exercise 6B (page 114)

1 a $2.09\,\mathrm{rad\,s^{-1}}$

  b $3.14\,\mathrm{m\,s^{-1}}$

  c $6.58\,\mathrm{m\,s^{-2}}$

2 $8\,\mathrm{N}; 7.75\,\mathrm{m\,s^{-1}}, 12.9\,\mathrm{rad\,s^{-1}}$

3 $0.24\,\mathrm{m}$

4 $0.133\,\mathrm{N}$

5 $0.781\,\mathrm{kg}$

6 0.375

7 $7 \times 10^4\,\mathrm{N}$

8 $27\,\mathrm{N}$

9 11.0 revolutions per minute

10 $\dfrac{ma}{M + m}$

11 $36\,200\,\mathrm{km}$

12 $0.034\,\mathrm{m\,s^{-2}}$

Exercise 6C (page 117)

1 a $1.19\,\mathrm{m\,s^{-1}}$

  b $8.49\,\mathrm{N}$

2 a 27.1°

   b 3.53 rad s⁻¹

3 a 11.7 N, 6.98 rad s⁻¹

   b 8 N, 5.77

   c 0.96 N, 3.52 N

4 i 1.86 m s⁻¹

  ii $\omega = 5.37$ rad s⁻¹

5 $\omega = 6(0.00)$ rad s⁻¹

6 i 4.1(0) m s⁻¹

  ii 5.59 rad s⁻¹

7 i $\omega = 6$, $T = 4.32$ N

  ii a $r_P = 0.6$ m, $r_Q = 0.4$ m

    b $v_P = 1.2$ m s⁻¹, $v_Q = 0.8$ m s⁻¹

8 36.9, 25 000 N; 67.3, 51 800 N

Miscellaneous exercise 6 (page 120)

1 1.70 N

2 $\omega = 0.5$ rad s⁻¹

3 i 3.6 N

  ii 0.882 N

4 i 3 N

  ii 3 m s⁻¹

5 i 1.5 N

  ii 1.73 m s⁻¹

6 ii 2.31 m s⁻¹

  iii 1.09 s

7 i 1.94 N

  ii 1.01 N

  iii 4.32

8 i 2.88 N

  ii 0.45 J

9 i $\omega = 8.19$, KE $= 0.402$ J

  ii $T = 6.71$ N, $\omega = 10.6$

10 2.98, 52.0 km h⁻¹

7 Geometrical methods

Exercise 7A (page 125)

1 a 1640, 72, 5.5 m

  b 1810, 96, 6.1 m

  c 1220, 125, 4.8 m

2 a 1400, 14.6, 10 m   **b** 1970, 40, 3.95 m

  c 1800, 58.7, 5.33 m   **d** 2750, 66.2, 6.09 m

3 2.08 m

4 100 N along the diameter AB

Exercise 7B (page 129)

1 41.6 N at 16.1° to the upward vertical, 11.5 N

2 480 N at 38.7° to the horizontal

3 1.71 m

4 a 16.4 N at 13.2° to the upward vertical

  b 16.4 N at 13.2° to the upward vertical in the opposite sense to part **a**

5 9 m, 373 N, 236 N

6 50°; 61.3 N, 51.4 N

7 50°

8 37.9

Exercise 7C (page 132)

1 a 43.6 N

  b 23.4°

  c 0.433

2 a 1147 N

  b 582 N

3 b 0.424

4 a 47.6 N

  b 9.1° to the upward vertical

  c 0.599

5 33.6° to the upward vertical; 18.6°

6 a 35.0°

  b 11.9

  c 20.0

7 a 755 N at 14.4° to the upward vertical,

   i 200 N at 20° below the horizontal in to the tree,

  ii 755 N at 20° to the downward vertical

  b 752 N at 22.0° to the upward vertical,

   i 300 N at 22.0° below the horizontal in to the tree,

  ii 752 N at 14.5° to the downward vertical.

The force exerted on the man by the ground must act upwards; $P > 800/\cos 70°$ is inconsistent with this.

Exercise 7D (page 135)

1 0.289; 0.520

2 0.346

3 0.940

4 237; 0.297

5 More than 0.276 m from the back of the sledge

6 **a** 25 N

   b Greater than 0.174

Miscellaneous exercise 7 (page 136)

1 On AD 2 m from A, 100 N at 36.9° to DA

2 7.31 m from the centre of mass, 40°

3 36.9

4 0.335

5 **ii** $\frac{11}{48}$

6 19.1

8 587 N

8 Centres of mass of special shapes

Exercise 8A (page 143)

1 16.4 cm

2 **II** 15.3°

4 **ii** 0.117 m

5 0.1 cm

6 (5.56, 5.08)

8 0.866

9 0.453 m

10 **a** P   **b** R

    c Q   **d** S

11 10.4°

12 $2kr\alpha, 2kr(\pi-\alpha); \dfrac{r\sin\alpha}{\alpha}, \dfrac{r\sin\alpha}{\pi-\alpha}$

Exercise 8B (page 150)

1 **a** Cylindrical part in contact

    b Hemispherical part in contact

2 10.25 cm, 10.29 cm from the two sides; no

3 66.3°

4 2.46 cm, 4.95 cm

5 $0.316l$

6 11.2°

7 3.85°

Miscellaneous exercise 8 (page 152)

2 **ii** 13(.0) N

7 1.09 W at 23.0° to the vertical

9 The vase topples without first overflowing.

10 7.75 cm

11 1.60 cm

12 18.45

13 4.24 cm

9 Linear motion with variable forces

Exercise 9A (page 161)

1 $2\,\text{m s}^{-1}$, $2\frac{2}{3}\,\text{m}$

2 $80\,\text{m s}^{-1}$, $600\,\text{m}$

3 $(2-2\cos 2t)\,\text{m s}^{-1}$, $(2t-\sin 2t)\,\text{m}$

4 $75, 63\,\text{m}$

5 $\dfrac{K}{mc}\left(1-e^{-ct}\right), \dfrac{K}{mc^2}\left(ct+e^{-ct}-1\right)$

6 $5.36\,\text{m s}^{-1}$, $132\,\text{m}$

7 6.4 s, 129 m; speed never becomes zero with parachute and air brakes alone

8 118 s, 1.15 km

9 1.54 s, 3.88 m; 1.5 s, 3.75 m

10 6 s, 65 m, $10.8\,\text{m s}^{-1}$; 8.20 s, 91.5 m, $11.2\,\text{m s}^{-1}$

11 $\displaystyle\int_{0}^{80} \dfrac{500\,000v}{25\,000\,000 - 0.8v^3}dv$, 64 s

12 $54.9\,\text{s}; \displaystyle\int_{0}^{50\ln 3} 20\,\dfrac{e^{0.02t}-1}{e^{0.02t}+1}dt$, 288 m

Exercise 9B (page 169)

1 **a** $v^2 = 16x+1$ **b** $v^2 = 36 - 4x^2$

    c $v = -2x$ **d** $v = -x^2$

2 **a** $v = 1 + x^2$

    i $2x(1+x^2)$

    ii $2v\sqrt{v-1}$

    b $v = 1 + \frac{1}{4}x$

    i $\frac{1}{4}\left(1+\frac{1}{4}x\right)$

    ii $\frac{1}{4}v$

    c $v^2 = x - \frac{1}{2}x^2$

    i $\frac{1}{2}\left(1-\frac{1}{2}x\right)$

    ii $\pm\frac{1}{2}\sqrt{1-v^2}$

d $v^2 = 4 + x^2$

    i x

    ii $\sqrt{v^2 - 4}$

3 $v = 8 - 0.02x$

4   129 m

5   1.15 km

6   288 m

7   79.8 m

8   3.44 km

9   204 m

10   3.45 km

11 $x = 10\dfrac{e^{2t} - 1}{e^{2t} + 1}; \dfrac{40e^{2t}}{\left(e^{2t} + 1\right)^2}$

12 $x = \frac{1}{4}qt^2 + \sqrt{pt}; p = u^2, q = 2a$

13 $v^2 = n^2\left(c^2 - x^2\right), -n^2 x$

14   23.0 J

15   24 J

Exercise 9C (page 175)

1   **a**   50 m s^{-1}

    b   44.7 m s^{-1}

    c   7.19 s

2   **a**   1.81 km

    b   225 m s^{-1}

    c   14.1 s

3   **a**   22.3 m

    b   17.6 m s^{-1}

    c   4.26 s

4   **a**   26.4 m

    b   21.2 m s^{-1}

    c   4.59 s

Miscellaneous exercise 9 (page 175)

1   **ii**   5.33 m s^{-1}

2   **i** $x = 2.5, v = 6.12$ m s^{-1}

    ii -15 m s^{-2}

3   **ii**   5.35 m s^{-1}

4   **ii**   12.9

5   **i** $v = 50 - 50e^{-0.2t}$

    ii   17.6

6   **i**   4.08 m s^{-1}

    ii   6.67 m

7   **i**   0.738 s

    ii   0.39 m

8   **ii** $v = x^{\frac{2}{3}}$

    iii   3

9   **i**   2.75 s

    ii   4.51 m

10   **i** $0.25v\dfrac{dv}{dx} = -(5 - x)$

    ii $x = 5(1 - e^{-2t})$

10 Strategies for solving problems

Exercise 10A (page 183)

1   **a**   52.0 N at 30° to the vertical

    b   39.7 N at 19.1° to the vertical

    c   30 N vertical

2   **a**   693 N

    b   400 N at 60° to the vertical

3   **i**   0.955 N

    ii   2.22 N

4   0.372

Exercise 10B (page 189)

1   **a**   41.2

    b   41.3 m s^{-1}

2 $\tan^{-1}\frac{3}{4}$

    a   3.16 s

    b   25.3 m s^{-1}

    c   20.5 m

3   4.63 rad s^{-1}

4   20.8 m s^{-1}; 614 N

5 $\frac{1}{2}ml\Omega^2 - \frac{1}{2}mg\sec\theta$

Miscellaneous exercise 10 (page 191)

1 $2\frac{2}{3}$ m from O

2   **i**   5.25

3   102 m s^{-1} at 47.5° to the horizontal

4   **ii** $\lambda = 700$ N

    iii $F = 13.5$ N, $\alpha = 42.(0)°$ (with horizontal)

6   **a** $y = \dfrac{2\pi^2}{gT^2}x^2 + k$

    b $y = \dfrac{V^2}{g}\ln x + k$

    c $y = -\dfrac{c}{gx} + k$

Revision exercise 2 (page 193)

1 139, 102 N at 40.9° to the horizontal

2 a 103 m

b 128 m; air resistance and aerodynamic lift

3 3.54 rad s^{-1}

4 8.5 s

5 a 49.0 N

b 110 N at 72° to the horizontal

6 32.9 cm from B; 1.43°

7 37.3

8 ii b $\mu < \frac{7}{5}$

9 12.5 N

10 ii $\omega = 4.58$ rad s^{-1}

iii 2.04 m s^{-1} or 1.6 m s^{-1}

iv 2 m s^{-1}

11 a if $\alpha > 54.7$

b if $\alpha < 54.7$

12 ii 59.4 m

13 26.6°; 93.4 grams

15 $\frac{7}{33}$ cm

16 ii $v = 30e^{-t} - 10$, (until $0 = 30e^{-t} \Rightarrow$ valid for $0 \le t \le \ln 3$)

iii -30 m s^{-2}

17 $\tan^{-1}(3 + 2\mu)$

18 i $x = Vt \cos 60$, $y = Vt \sin 60 - \dfrac{gt^2}{2}$

ii 29.7

iii 55.3° downwards from the horizontal

19 9406 N, 6.18 m s^{-2}, 166

20 $\theta \ge 0.224$

Practice Exam-style papers

Practice Exam-style paper 1 (page 198)

1 $\frac{8}{25}W$

2 23 N

3 4

4 ii $\sqrt{20}$

5 i $mv\dfrac{dv}{dx} = kx^{-\frac{3}{2}}$

ii 240 000 ms^{-2}

6 6.88 s

7 ii 12 000

iii a 53.1°　　　　**b** 11.6°

Practice Exam-style paper 2 (page 200)

1 i $\dfrac{40}{3\pi}$ cm

ii 23.0°

2 ii $\dfrac{10}{\sqrt{3}} \approx 5.77$

3 $\frac{4}{9}$ m

4 i 1.75 m

ii $\sqrt{20} \approx 4.47$ m s^{-1}

5 i $\sqrt{200} \approx 14.1$

ii $\dfrac{25}{\sqrt{2}} \approx 17.7$; 22.6 m s^{-1}, 38.7° with the horizontal

6 i 3.125

iii $v = 80\sqrt{1 - e^{-\frac{1}{120}t}}$

7 i a $\frac{2}{3}W \cos\alpha$

Index

The page numbers refer to the first mention of each term, or the box if there is one.

Formulae

Uniform acceleration formulae

$$v = u + at, \qquad s = \tfrac{1}{2}(u+v)t, \qquad s = ut + \tfrac{1}{2}at^2, \qquad v^2 = u^2 + 2as$$

Motion of a projectile

Equation of trajectory is:

$$y = x\tan\theta - \frac{gx^2}{2V^2\cos^2\theta}$$

Elastic strings and springs

$$T = \frac{\lambda x}{l}, \quad E = \frac{\lambda x^2}{2l}$$

Motion in a circle

For uniform circular motion, the acceleration is directed towards the centre and has magnitude

$$\omega^2 r \text{ or } \frac{v^2}{r}$$

Centres of mass of uniform bodies

Triangular lamina: $\tfrac{2}{3}$ along median from vertex

Solid hemisphere of radius $r : \tfrac{3}{8}r$ from centre

Hemispherical shell of radius $r : \tfrac{1}{2}r$ from centre

Circular arc of radius r and angle at centre 2α: $\dfrac{r\sin\alpha}{\alpha}$ from centre

Circular sector of radius r and angle 2α: $\dfrac{2r\sin\alpha}{3\alpha}$ from centre

Solid cone or pyramid of height $h : \tfrac{3}{4}h$ from vertex